DATE DUE

FEB 1 0 2007			
GAYLORD			PRINTED IN U.S.A.

Ontologically Controlled Autonomous Systems: Principles, Operations and Architecture

Ontologically Controlled Autonomous Systems: Principles, Operations and Architecture

by

George A. Fodor
ABB Industrial Systems AB, Sweden
Western Michigan University, USA

KLUWER ACADEMIC PUBLISHERS
Boston / Dordrecht / London

Distributors for North America:
Kluwer Academic Publishers
101 Philip Drive
Assinippi Park
Norwell, Massachusetts 02061 USA

Distributors for all other countries:
Kluwer Academic Publishers Group
Distribution Centre
Post Office Box 322
3300 AH Dordrecht, THE NETHERLANDS

Library of Congress Cataloging-in-Publication Data

A C.I.P. Catalogue record for this book is available
from the Library of Congress.

CONTENTS

LIST OF FIGURES

FOREWORD

Kevin M. Passino

When confronted with a control problem for complicated physical process, a control engineer usually follows a predetermined design procedure. This procedure often begins with the engineer seeking to understand the process and the primary control objectives. A simple example of a control problem is an automobile "cruise control" that provides the automobile with the capability of regulating its own speed at a driver-specified set-point (e.g., 55 mph). One solution to the automotive cruise control problem involves adding an electronic controller that can sense the speed of the vehicle via the speedometer and actuate the throttle position so as to regulate the vehicle speed at the driver-specified value. Such speed regulation must be accurate even if there are road grade changes, head-winds, or variations in the number of passengers in the automobile.

After gaining an intuitive understanding of the plant's dynamics and establishing the design objectives, the control engineer typically solves the cruise control problem by using an established design procedure. In particular, this control engineering design methodology involves:

1. Modeling/understanding the plant,

2. Construction of a controller to meet specifications (such as stability, rise-time, overshoot, and steady state error),

3. Analysis to make sure that the system will meet the performance objectives (e.g., we might use mathematical, simulation-based, or experimental analysis), and

4. Iterating on the design until it is possible to "commission" the control system.

This methodology has been used for a variety of applications including: the cruise control problem described above, other automotive systems (e.g., brakes and transmission), aircraft control, satellite altitude control, automated highway systems, ship steering, robots, and process control to name a few.

There are also control problems that have a more "sequential nature" where it is more convenient to represent the plant with a "discrete event system mode" (e.g., an automata or Petri net). There is a similar need for developing feedback control systems for such problems as there is for applications listed above, and a similar methodology to the above one is used. For example, the control (scheduling) of flexible manufacturing systems is of significant importance for enhancing product throughput and in such problems one is concerned with issues such as stability (e.g., guaranteeing that the queues of parts waiting to be processed will not grow without bound).

Unfortunately, while the control problems that are of sequential nature can be every bit as important as those of continuous nature there has not been as much progress in their solution. Note that the control problem for linear plants is largely solved (consider the use of classical techniques and the wide use of proportional-integral-derivative (PID) control in industry. The control of nonlinear and uncertain systems, is the focus of a significant amount of current research (e.g., robust nonlinear and adaptive control). The area of discrete event systems, where some some sequential control problems are studied, is receiving more attention but has had very little impact on the solution of practical industry problems (except, perhaps, if we consider work in scheduling to be a control problem), especially in impacting how problems are solved on a day-to-day basis on the factory floor.

Essentially, there is a large "theory-practice gap", where the theory that is being developed for discrete event systems is ignoring many practically important issues that are encountered when one really seeks to implement a sequential control algorithm. Basically, it is a fact that programmable controllers (PC) or programmable logic controllers (PLC) implement many industrial control systems yet very little attention has been paid to problems associated with PCs/PLCs in the academic research community. This book seeks to fill in the theory-practice gap by providing a theory for a set of very practical sequential control problems that are solved with PCs/PLCs.

Overall, there is a set of complex decision-making strategies that are relevant to the very practical problems studied in this book. Some of these come from the areas of discrete event systems and hierarchical control, while others come from artificial intelligence (AI) planning systems. It is the case, however, that not nearly as much academic literature is relevant to this book as is usually the case. Basically, the approach in this book is to identify and solve a problem that is encountered in many applications that were encountered in industry. Due to the practical motivation for the very problem being studied, the solutions become particularly relevant to industry and therefore of significant interest.

Basically, the problem being studied is that of supervisory control of programmable controllers for industrial applications. The work builds on the current industrial practice of using PLCs and PCs and provides strategies for detecting, identifying, and recovery from "problematic control situations" such as:

- Loss of communication with a sensor of the plant

- Desynchronization from the plant

- Deadlock

- When the controller thinks that the plant is in one state when really it is in another

While it is relatively clear how to identify and recover from problematic control situations for some simple plants, it can be very difficult to detect and recover from these in very complex real-world industrial control applications where there are many unpredictable causes for these. In addition, there may be a parallel development of parallel PCs/PLCs, or hierarchies and distribution of controllers for autonomous robots/vehicles or large industrial control systems. These situations can provide especially complex and challenging problematic control situations.

The basic question addressed in this book is: What should the PC do to respond to a problematic control situation? This book answers this question by providing methods that will automatically seek an achievable and appropriate goal for the controller when a problematic control situation occurs. The identification and recovery operations basically involve two steps: (i) a "state synchronization operation" determines what controller state corresponds to the current plant state and (ii) a "goal seeking operation" finds the highest priority goal and a control sequence to achieve it.

Several methodologies are provided for programmable controller supervision and there are many examples that are woven throughout the book to illustrate the application of the theory.

Overall, the issues studied in this book provide challenging theoretical problems that one would hope future research would address in even more detail, and for additional applications. At the same time the solutions that are provided here have significant practical ramifications and I would expect that engineers in industry will find them useful in solving real-world problems.

Kevin M. Passino
Columbus, Ohio, USA

This book presents the main principles, operations and architecture involved in the design of a novel type of supervisory controller called an *ontological controller*. An ontological controller can be used to supervise any type of controller, however its intended applications are industrial-strength complex autonomous control systems using advanced programmable controllers. An ontological controller supervises a *programmable controller* in order to:

- Detect dynamically when the programmable controller is in a problematic control situation due to a violation of *ontological assumptions* and thus, unable to achieve a pre-specified control goal (i.e. the *identification* operation), and

- When possible, move the programmable controller in such a state from which it can regain its control and eventually achieve the pre-specified control goal in spite of the previous violation of ontological assumptions (i.e. the *recovery* operation).

The ontological assumptions are essential for the correctness of the control algorithm of the programmable controller, but are implicit in it. A programmable controller succeeds in achieving a pre-specified control goal only if the ontological assumptions are not violated during the execution of its control algorithm. Since the ontological assumptions are not explicitly represented in the control algorithm, the programmable controller itself is not "aware" of them and violations of these cannot be detected by it.

A control paradigm which can be used to provide a proof that the ontological assumptions are violated during the execution of the control algorithm, or that they were simply incorrect already during its design is called *ontological control*.

1

INTRODUCTION

1.1 Scope

Today many industries secure essential competitive advantages by deploying advanced automation and control equipment. The following application areas depend critically on the proper design and implementation of control equipment:

- Large process industries such as chemical, petrochemical, pharmaceutical, food, pulp and paper;

- Heavy industries such as metallurgy, building materials and mining;

- Electrical power generation: land-based, marine and off-shore platforms;

- Automotive industries;

- Unmanned and autonomous applications: unmanned marine vessels, missiles, autonomous robots and unmanned spacecraft.

Both in the industry and academia an intensive research effort aims to find new control principles and new ways to design advanced control systems. Industry requires new control systems because of emerging technologies, the growth of a global economy and requirements for environmental protection. The following factors are particularly relevant for the subject of this book:

- *Managing complexity.* Many industrial applications are large, diverse and often geographically distributed. Operators and application programmers expect that the modeling, programming and supervisory tools can assist for a better understanding of the dynamics and structural properties of the application. The existence of advanced supervisory tools reduces the commissioning and maintenance time, and increases operational safety.

- *Application-independent fault detection and recovery.* Nearly all industrial projects in discrete control utilize off-the-shelf programmable controllers (PCs). The companies integrating these PCs have increased expectations that application-independent fault detection and fault recovery functions are built into PCs. PC language standards such as IEC 1131-3 [12] define a number of fault detection principles for individual PCs. However formal verification requirements not covered today by PC standards as well as the requirements for complex systems integrating many PCs, are steadily increasing.

- *Autonomy and resilience to faults.* To qualify for operational continuity requirements, most applications need a high degree of autonomy. Complex applications and a fast rate of technology transfer in a global economy lead to control systems which consist of a vast number of types and makes of equipment. Yet, such a heterogeneous system is required to act consistently even in cases of unforeseen situations such as partial equipment failure, system start-up in arbitrary states, or manual and automated operation modes for subparts of the equipment. Due to safety requirements, a substantial degree of autonomy is built-in even when human operators supervise the system.

- *Flexible manufacturing.* Flexible manufacturing (production in small batches) requires that control systems can be reconfigured often. A typical sequential control application which requires reconfiguration is the *flexible manufacturing system* (FMS) which consists of a number of robots that assemble or manufacture a product in co-operation with each other. Increased reconfiguration means that operations such as fault detection, factory acceptance tests (FAT) and commissioning require more time to perform.

The controllers considered in this book fall in the category of so called *discrete, logic* or *procedural controllers* for which state-based or logical type of decision is dominant. Although most applications do have continuous type of control, the continuous control part is normally subsumed to the overall sequential logic and the largest programming effort and complexity issues are related to the sequential part.

The scope of this book relative to the research activity, existing applications and the technology in advanced discrete control can be described as follows:

- The programmable controller technology improves continuously. A typical control system has now computing power, communication and programming capacity many times the size of the first generation of programmable controllers. However, the programming languages, the modeling and the source code formal verification tools used today with programmable controllers depart very little from those used with earlier systems. The hardware-related advances in PC design have been confined mostly to quantitative aspects such as speed, program size and graphical presentation. For example, the standard IEC 1131-3 for programmable controllers [12] defines the most popular PC programming languages and their syntactical and semantical verification principles. Still the standard has limits regarding program synthesis, program transformation and advanced program verification principles.

- On the application side, several advanced control architectures and programming techniques have been developed by application software houses and programmable controller manufacturers. Most of these results are not available in publications. Even when published, (e.g., [1], [17], [6]), the results are related to specific applications and the benefits of the reported results for a general autonomous control architecture are often hard to estimate due to lack of reference cases and lack of uniform terminology.

- Research in computer sciences and artificial intelligence (AI) has made available many results in fields such as real-time planning, automatic program verification and program synthesis. These results are based on tools such as temporal logic, discrete-event systems, or process algebra. Still, the industry has implemented rarely and in small scale these results in domestic control architectures (an exception is perhaps the *Grafcet* language, inspired from Petri nets [7]).

- Due to the differences between technology, applications and research shown above, the problems related to complex autonomous control systems are not so well defined as say, in continuous control. This situation is well acknowledged in literature, e.g., [20], [3]. It is often the case that the research effort in industry and academia are following different objectives.

Recent research seems to indicate a departure from regarding control primarily as the operation of reaching goals by movements in a state space. Instead, the focus is on the operations which maintain the internal environment of a system [24]. As shown in Section 3.4.4, this orientation is closer to the current programming experience with advanced programmable controllers.

1.2 Autonomous Complex Control Systems

Even simple applications can use the control principle presented in this book. However the main intended applications for ontological control are industrial-strength complex autonomous control systems. These systems have a number of features normally attributed to "intelligent", "autonomous" or "hybrid" systems. The following sections outline the main features assumed in this book for the control system.

1.2.1 Programmable controllers and complex control systems

In industry it is largely assumed that in the next few years advanced control systems for industrial applications will still be based on programmable controllers. This assumption is based on the size of the already installed application base and the existence of established programming languages, standards, safety rules and application libraries. At a basic level, programmable controllers perform traditional continuous and discrete type of control by sensing and then acting on the controlled plant or environment. Furthermore, the system may have a supervisory or more "intelligent" type of control which is enabled on-line when the basic level fails to act properly. The reasons for introducing a *control architecture* — that is, organizing the control on hierarchical levels — will be analyzed and when possible, proved.

1.2.2 Hybrid control

Typical complex control applications include both traditional continuous control and sequential control. Hybrid control is the study of control of continuous dynamic processes by discrete-state sequential machines [2]. For a typical programmable controller, the traditional continuous control consists of dedicated hardware or of software libraries. Each continuous part works with the overall sequential control by way of signals indicating stable or relevant operation modes in the continuous domain. Almost all manufacturers of programmable controllers use ad-hoc separation principles between the continuous and sequential part, though results on separation principles such as [5] are available.

Normally the size of the sequential part of the application program is much larger than the size of the continuous part [4]. The reason is the sequential nature of applications in process-related industries (controllers for such applications are called *procedural controllers* [20]) and the huge number the logical operations required, such as safety checks, interlocks and reconfigurations.

1.2.3 Control architecture and computational complexity

Research in Artificial Intelligence revealed early that advanced systems, such as those designed to meet unexpected situations or to make plan synthesis, must deal with the problem of combinatorial explosion. Each research field closed to real-time control tries to solve this problem, e.g., intelligent systems [13], real-time planning [23] or synthesis of discrete-event systems [20]. For complex industrial systems with firm real-time requirements, algorithms implying combinatorial explosion may not be usable. The trade-off used in practical applications to achieve both advanced features and acceptable computational complexity is to use the control architecture as a mean to avoid combinatorial explosion. A typical architecture has at a basic level a reactive type of control with a high event rate (lowest time granularity). At this level all the relevant control situations are treated with the highest possible speed using predefined control actions. The computational complexity at this level is low. All the control situations which cannot be covered at this basic level (e.g., due to lack of some required global knowledge, lack of proper local control action or lack of awareness for the particular case), are handled at a higher control level. The price for using a hierarchical architecture is an inherent lower event rate at the higher level [18]. Moreover the same state or procedure at the higher level may be mapped to several unexpected or unobserved states at the basic level, i.e. the system may not have specific response to certain unexpected events. However the complexity is reduced since not all the possible cases need to be considered at the lower level and several cases with the same relevance are treated identically at the higher level. As with other features described in this book, those architectures are relevant which can reduce complexity independently of a particular application.

1.2.4 Dynamic Complexity

Complexity in industrial applications is caused primarily by the intricated time development of the sequential control part. Normally, a control program is executing sequences of actions in order to move controlled objects into predefined states in a certain (partial) order. However, this operation can be disrupted by different events (expected or unexpected) such as operator commands, disturbances or equipment error. After such an event, the controller makes a *reevaluation* of its future actions based on (*i*) what object states have been already reached, (*ii*) what object states have been pursued at the time the disturbing event has occurred and (*iii*) what object states can be possibly reached after the occurrence of the disturbing event. Even for applications controlling few objects the reevaluation operation can generate considerable dynamic complexity.

1.2.5 Autonomy

The report of the Task Force on Intelligent Systems (Antsaklis et. al. [2]) gives several definitions for Autonomous Systems, such as:

- "...an autonomous system ... designs control laws to meet well-defined control objectives";
- "...autonomous system is when the components, control laws, plant models are not completely known";
- "...an autonomous system ... uses procedures of focusing attention, combinatorial search, generalization, which are applied to the input information to produce the output".

Unlike the uniform computing environment in research laboratories, large control systems are being made autonomous by application programs executed on heterogeneous programmable controllers. Thus a definition for what is an autonomous system based on *focus of attention, generalization,* or *combinatorial search* may be difficult to verify due to implementation peculiarities and programming language characteristics. Instead, we describe an autonomous system in terms which are application-independent, indicating the control architecture which gives the autonomous behavior.

As already described, certain events can interrupt a programmable controller from the execution of a predefined sequence of control actions. In this case, the programmable controller may have mechanisms to reevaluate and perhaps change or re-order the priority of its future goals. This operation is based on what goals the programmable controller has already achieved, what control sequence has been pursued at the time the event has occurred and what alternative goals can be followed given the current situation. The reevaluation performs a so called *goal seeking operation* with the following properties:

1. The algorithm can determine what is the current control situation (the so called *state synchronization operation*);

2. The algorithm can determine which is the control sequence towards the goal with the highest priority in the current control situation;

3. The algorithm is executed off-line, i.e. all the required input data is already available when the algorithm starts;

4. The algorithm is not specific to a certain application; (that is, it might be implemented in the operating system of a programmable controller rather than being re-created with each application).

Relative to the definitions of an autonomous system above, a control system equipped with a goal seeking operation is autonomous in the following sense:

- The control system makes an off-line reevaluation of its control objectives when it encounters a disturbing situation which has not been considered at design time. In other words, the goal seeking will act when components, plant models or control laws are not completely known.

- The goal seeking operation changes the current *"focus of attention"* — i.e., the currently pursued sequence and corresponding goal state, — from a goal the control system cannot achieve to a goal it can achieve and which has the highest priority in the given situation;

- The generalization or the abstraction level of the goal seeking algorithm is a matter of how complex the algorithm is and does not influence the property of the system of being autonomous;

- A programmable controller which can be equipped with, or which already has a goal seeking operation with the properties above can be termed *autonomous* independently of its particular application or goal seeking algorithm since the autonomy originates from the architecture.

1.2.6 Control Intelligence

This section describes a property for complex control systems which fall in the class of so called *Control Intelligence* [2]. The relation between Ontological Control and Control Intelligence, can be understood by analyzing the main concepts shown in Figure 1. This conceptual space, familiar to workers in computer science and real-time control, consists of the following three parts:

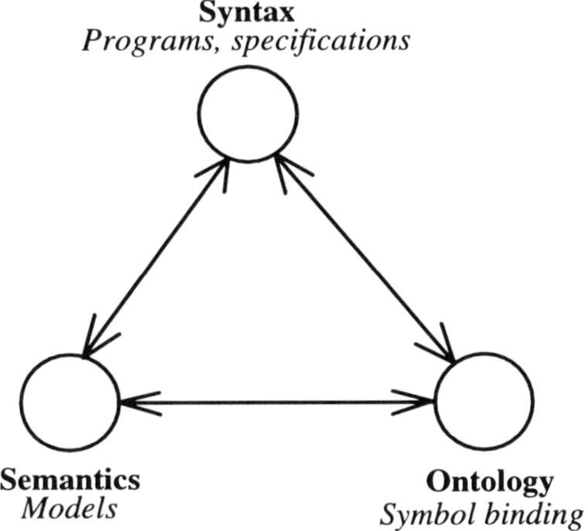

Figure 1 Conceptual Space.

Syntax refers to the collection of symbolic or textual information used with control applications. Examples are computer programs and formal specifications.

Semantics (Models) refers to the models underlying the syntactical representation, i.e. the interpretation of each syntactical symbols and combination of symbols. An example is the interpretation of the symbols of an application program.

Ontology refers to the relation between how the state symbols are bound with sensed real-time data relative to the modeled (or expected) effects of the control actions. If in a certain control state a controller performs an action such that the expected effects of this action do occur upon the controlled plant or environment, then the ontology is right for the given application domain. For example if an electrical current is above a given critical value, the controller supervising the electrical network may execute a breaker *open* command. Then the electrical current is expected to become zero, i.e. the variable denoting electrical current is expected to change in a predictable way and all the control states in which the variable *electrical current* appears are expected to materialize accordingly. That means the sequences of control states reflect certain ontology or theory about the world outside the controller. Let us assume that the breaker has a built-in safety lock which protects the contacts from breaking a high current. This lock disables input *open* commands when the current is too high. Moreover, let us assume that the designer of the control program was not aware at design time of this safety feature of the breaker. Then, when the current is high, the controller issues the *open* command which has no effect and the electrical current will continue to be high. That means the control algorithm has a wrong ontology relative to the breaker, or the breaker presents violations of the ontological assumptions relative to the control algorithm.

The theory presented in this book analyzes the relation between ontology semantics and syntax, such that a complex autonomous system can detect and possibly make an unatended recovery after violations of the ontological assumptions. In the presence of violations of the ontological assumptions, a control state does not fulfill the expectations built into the control program. That means some state is true when interpreted with sensed signals, however this state it is not the intended one for the given situation (the state symbol is incorrectly bound) and the control system performs unwarranted control actions. After executing a control action that is not the intended one for the current situation, the control system normally loses control upon its environment. An ontological controller uses the following steps to make a programmable controller regain its control upon the environment and still achieve the initial goals:

1. The ontological controller identifies that an ontological violation has occurred. The programmable controller can continue to act since as shown later on, in the presence of ontological violations the programmable controller executes an infinite control cycle.

2. The ontological controller performs syntactical operations on a copy of the state set of the PC. It identifies new controller states that are ontologically right and remove culprit state(s) which have violations of the ontological assumptions. The symbols of the state set are changed accordingly.

3. The ontological controller changes on-line the state set of the PC. The infinite control cycle is broken and the PC regains its control over the plant.

As it turns out, the identification operation is the most difficult one since ontological violations are very hard to separate from other unexpected events that may occur in a complex environment. Thus the largest part of this book deals with finding the syntactical conditions under which ontological violations can be identified.

The autonomous behavior realized by the goal seeking operation and the ontological control level, meet some of the definitions for *control intelligence* outlined in [2]: autonomy, reconfiguration, reformulation of assignment, variable symbol binding and alternative paths.

1.2.7 Finite state machine models

Operations such as synthesis, verification and modeling of sequential control systems are often performed on a formal representation based on finite state machines (FSM). Restricting our attention to one of the most used formalism — the theory of discrete event systems (DES) of Ramadge and Wonham [19] — (a summary with other references to FSM approaches in can be found in [20]), the notation used with this theory is:

$$G = (Q, \Sigma, \delta, q_0, Q_m)$$

where:

G = is an automaton or language generator. The language generated by G, denoted L(G), is a regular language (a *regular language* is one generated by a finite state machine);

Q = set of all possible states;

$\Sigma = (\sigma_1, \sigma_2, \dots)$ = is the set of all events. Represents the alphabet of the language. A trajectory of events is a string in the language. The set of all possible strings is denoted Σ^* which includes the empty string ε representing no event;

$\delta : \Sigma \times Q \to Q$ is the state transition function;

q_0 = initial state;

Q_m = set of marked states;

The set of G-allowed strings in the language, L(G), is generated by the state transition function δ the output being certain set of strings of events σ allowed by δ. The set of all possible state trajectories in the language is very large and the purpose of the synthesis or of the verification operation is to restrict possible trajectories to those which satisfy certain conditions or specifications. Most used frameworks for expressiong these restrictions are the linear temporal logic (LTL) and/or the supervisory control theory (Ramadge and Wonham [25]). Some typical examples are in (Thistle and Wonham, [21], Lin, Ionescu [14]) and Lin [15]). The controller synthesis or verification operation gives a subset $L_v(G) \subseteq L(G)$ of allowed event trajectories.

Let us assume that a synthesis operation for a controller generates the subset $L_v(G)$ such that it satisfies a given plant specification and closed loop conditions (for a detailed example see e.g., (Thistle and Wonham, [21])). Since $L_v(G)$ is available, it can be implemented in an actual controller. During the execution of the control program, the states in $L_v(G)$ can in principle be followed as they materialize in real-time. Thus, the event trajectories in $L_v(G)$ can be inspected in two instances: *(i)* at generation time and *(ii)* at execution time. Thus, we can distinguish a *generation model* and an *execution model* (Figure 2):

1. *The generation model.* The set of all allowed trajectories $L_v(G)$ can be generated in principle starting with the pair (q_0, ε) consisting of the initial state and the empty event, and constructing all the transitions allowed by δ and the specifications available. Note that the generation of $L_v(G)$ is based on the syntactical part given by G and possibly additional models or specifications. However, there is no ontology involved at this step.

2. *The execution model.* The execution model uses a fixed $L_v(G)$. The states and events represented syntactically in $L_v(G)$ are interpreted with real-time data. The purpose of the execution model is to bound states and events such that the ontology of the syntactical entities in $L_v(G)$ are satisfied. In other words the execution is successful if at any time instance there is no event such that the execution 'falls' outside $L_v(G)$.

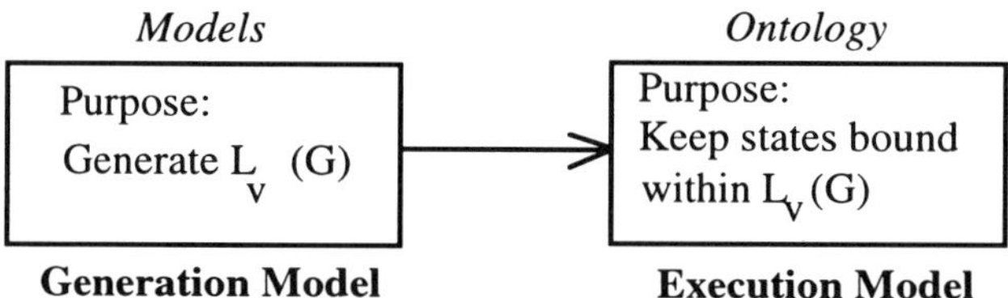

Figure 2 Generation and Execution Model.

In many cases the output of the generation model is held to be identical to the state set executed in real-time. However, the two models are not the same. On one side the execution model has its dynamics and its specific problems retated to that, as shown in the subsequent chapters. On the other side, the output of the generation model (a finite state machine, state set or other formal representation) often does not have the format or it is not intended to be directly executed. For example, the generation model of a discrete event system as described in [25] is not intended to be executed since it lacks the enforcement of control actions (control actions belong to the ontological part).

In this book we make a distinction between traditional disturbances that appear in the generation model and the problems related to the real-time execution of the state set of the execution model. We call the later as *problematic control situations.*

1.2.8 State representation

In this book we investigate problematic control situations related to the execution model. The generation model uses often search operations on the full *(event × state)* set product. By contrast, the execution model uses state sets and state transition functions which are fixed for a given application. In this context the state transition function δ is fixed and can be replaced by the index notation shown in Chapter 3. This notation uses the pair <pre-condition, control action> to represent the ontological and dynamical properties related to real-time execution of state machines. This representation is widely used even for synthesis and verification of real-time control systems ([9], [8], [22]). If needed, this formalism can be translated into any other one that is based on finite state machines.

The generation model and the execution model can be investigated even as a one step operation, as this is done in real-time planning. However the problem formulation and the proofs are formally complex when ontological violations are considered as well. This approach will not be considered in this book.

1.3 Ontological Control

Control theory studies how particular types of autonomous devices called *controllers* achieve pre-specified control objectives by conducting changes on a well-determined environment (e.g., a *plant* or a *process*). For this purpose, a controller executes a *control algorithm* which guarantees the achievement of the control objectives so that certain performance criteria, pre-specified constraints, or optimality requirements are met in spite of the possible occurrence of a number of predetermined types of disturbances. A control algorithm can be confined to just conventional linear or non-linear feedback control, or to just sequential control, or may involve both types of control which is the case with modern programmable controllers.

Independently of its nature, the control algorithm of a given controller is designed and optimized using an abstracted and formalized representation (a *model*) of all the objects under direct control (plant, actuators, sensors) as well as of the environment in which they are embedded. Besides the model, the algorithm may have incorporated a number of heuristics representing the unmodelled properties of the objects under control. In large industrial process control applications the environment is extremely complex and involves a hierarchy of other controllers, operator stations, local networks, etc., and some of these may interact in one way or another with the given controller.

However, when deriving a model, certain *modeling assumptions* are made which are **not** explicitly represented in the model and consequently they are not reflected in the control algorithm. For example, the equation $a=F/m$, is a formal model describing the acceleration of an abstract object as a function of its mass and a force applied to it. When this model is used to describe the behavior of a real physical object, the model variables are instantiated so that F becomes a mechanical force of certain size and range of variation, m becomes the real mass of the physical object, and a is perhaps the output of a particular sensor.

However, certain assumptions about the formal model and the physical object to which the model is to be applied are not made explicit during the instantiation phase. These assumptions are normally of two types: *model assumptions* and *model application assumptions*. For example, one model assumption may be that the mass of the abstract object **does not** create additional force when F is applied. A model application assumption may be that there are **no** electrical or magnetic forces acting on the real physical object besides F.

Assume now that the above model is used for the design of a control algorithm to control a real physical object whose mass **does** create an additional force when F is applied. In this case the model assumption from above is violated which in turn implies that the abstract model is not as applicable for this particular physical object. Consequently, the control algorithm may not achieve a control objective, such as to move the real physical object along a desired trajectory.

Now suppose that the model assumption is not violated by the real physical object, but during the execution of the control algorithm an additional magnetic force start acting on the real physical object. Thus, the model application assumption is violated (both a mechanical and a magnetic force are acting simultaneously). Then again, in the absence of a specialized control algorithm for this type of disturbance (e.g. robust control), the control objective may not be achieved due to deviations from the desired trajectory and instability. As observed in [13] both types of modeling assumptions, once made, are often neglected during the consequent design, optimization, and verification phases of a control algorithm.

In the case of modern programmable controllers for large industrial process control applications, the modeling assumptions in the control algorithms are extended by additional assumptions about the complex environment in which a given programmable controller is embedded. These additional assumptions refer to parts of the plant which the given controller is not directly controlling, but which interact with the plant under direct control; assumptions about the dynamic behavior of other controllers interacting with the given controller, etc.

We call *ontological assumptions*, the modeling assumptions and the additional assumptions about the environment which are implicit in a control algorithm, but which are essential for its validity. Thus, a control algorithm succeeds in achieving its control objectives **only if** the ontological assumptions implicit in it are not violated during the execution of this control algorithm. Since the controller executing a particular control algorithm is not 'aware' of the ontological assumptions made during the design of this control algorithm, it cannot detect itself a violation of the ontological assumptions.

In this context, this book describes a novel control paradigm which can provide a proof that a violation of ontological assumptions has occurred during the execution of the control algorithm. This control paradigm is called *ontological control*. The goal of ontological control is then to study the principles and operations involved in the design of a special class of supervisory controllers called *ontological controllers*. An ontological controller supervises a control algorithm in order to:

- Detect dynamically when the control algorithm under supervision encounters a situation in which the ontological assumptions for this particular control algorithm are violated, and

- When possible, to move the control algorithm in a state from which it can regain its control and achieve a control objective in spite of a previous violation of ontological assumptions.

In this book we consider the type of control algorithms used with modern programmable controllers. Thus, the nature of these control algorithms is sequential, but continuous feedback control elements may be embedded in such a sequential control algorithm. In this context, the goal of ontological control should not be confused with the problem of the verification of the properties of a sequential control algorithm such as reachability, dead-lock, etc. Furthermore, regarding the continuous feedback elements of a sequential control algorithm, the goal of ontological control is not to improve their performance in terms of stability, transient and steady state characteristics. Neither is ontological control to be considered as some form of an adaptive control, or expert supervisory control. The latter two types of control can vary adjustable system parameters (usually the controller parameters) so that satisfactory performance and stability can be obtained regardless of environmental changes.

1.4 Application Areas For Ontological Control

The following application areas depend critically on the proper design and implementation of ontological controllers:

- *Large industrial process control systems.* These are geographically distributed computer network systems with hierarchically organized execution of control functions. Such systems control the behavior of a large number of physically different objects (plants, processes, etc.) situated in diverse environments. Thus, the ontological assumptions underlying the different application programs are large in number and diversified in nature. Frequently, uncoordinated application-software design teams develop the software with rather local and possibly incorrect and/or contradictory, in the context of the whole system, assumptions about the controlled objects and their environments. The use of ontological control in large industrial process control systems is intended to minimize the risk of such globally, though locally correct and consistent, ontological assumptions. The targets for ontological controllers are the individual PCs placed in different nodes of the distributed system.

- *Autonomous agents,* such as unmanned marine vessels, missiles, and autonomous robots. Ontological control in the case of such autonomous agents is intended to detect violations of ontological assumptions as soon as the agent starts acting in an environment in which some of the ontological assumptions become violated due to changes in the environment and thus, the operation of an agent is rendered unsafe. Depending on the degree to which the ontological assumptions are violated, an ontological controller can correspondingly modify the sequential control algorithms embedded in such an agent so that it can act safely in the changed environment.

- *On-line diagnosis in process control.* In the case when the function of a control device fails a diagnostic program for this device has to discriminate between the case of a component failure having taken place and the case when changes in the environment in which the device is placed have violated some of the ontological assumptions underlying the control function of the device.

REFERENCES

[1]   Abu El Ata-Doss, S., Brunet, J. On-line expert supervision for process control. In the Proc. of 25-th Conf. on Decision and Control, Athens, Greece, Dec. 1986.

[2]   Antsaklis, P. et al. Report of the Task Force on Intelligent Control. IEEE Control Systems Society, Dec. 1993.

[3]   Antsaklis, P.J., Passino, K.M., *An Introduction to Intelligent and Autonomous Control.* Kluwer Academic Publishers, Boston/Dordrecht/London, 1994.

[4]   Benveniste, A., Åström, K. J., Caines, P.E., Cohen, G., Ljung, L., Varaiya, P. Facing the challenge of computer science in the industrial application of control: a joint IEEE CSS-IFAC project. In IEEE Transactions on Automatic Control vol. 38, No.7 July 1993.

[5]   William J. Bencze, Gene F. Franklin. A Separation Principle for Hybrid Control System Design. In IEEE Control Systems, 1995.

[6]   C. Charalambous, A.J. Conning. Distributed Sequence Control Utilising A High Level Sequencing Language in Conjection with PLC. IFAC AI in Real Time Control, Valencia, Spain 1994.

[7]   René David, Grafcet: A Powerful Tool for Specification of Logic Controllers. In IEEE Trans. on Control Systems Technology, vol. 3 no. 3, Sept. 1995, pp. 253-268.

[8]   Hatley, J.D., Pirbhai, I.A., *Strategies for Real-Time System Specification,* Dorst House Publishing, New York, 1987.

[9] Hendrickesn, C.S., Augmented State-Transition Diagrams for Reactive Software. In ACM SIGSOFT, Software Engineering Notes vol 14 no 6 Oct 1989 page 61-67.

[10] *** International Electrotechnical Commission. International Standard. Programmable Controllerrs Part 1: General Information. IEC 1131-1.

[11] *** International Electrotechnical Commission. International Standard. Programmable Controllerrs Part 2: Equipment Requirements and Tests. IEC 1131-2.

[12] *** International Electrotechnical Commission. International Standard. Programmable Controllerrs Part 3: Programming Languages. IEC 1131-3.

[13] Leitch, R.R. Modelling of complex dynamical systems. In IEE Proceedings vol. 134 Pt. D. No.4 July 1987.

[14] Lin, Jing-Yue, Ionescu, Dan. A generalized temporal logic approach for control problems of a class of nondeterministic discrete event systems. Proc. of the 29-th IEEE Conf. Decision and Control, Honolulu, Hawai, Dec. 5-7, 1990, pp3440-3445.

[15] Feng Lin, Analysis and Synthesis of Discrete Event Systems Using Temporal Logic. In Proceedings of the 1991 IEEE Intnl Symposium on Intelligent Control, 13-15 August 1991, Arington Virginia, USA.

[16] Fangzhen Lin, Shoham, Y. Provably Correct Theories of Action (preliminary report), in: National (U.S.) Conference on Artificial Intelligence, pages 349-352, 1991.

[17] R. Milne, C. Nicol, M. Ghallab, L. Trave-Massuyes, K. Bousson, CONTROL. Dousson, J. Quevedo, J. Aguilar, A. Guasch K. TIGER - real-time situation assessment of dynamic systems. In Intelligent Systems Engineering, Autumn 1994, pp. 103-124.

[18] Passino, K.M., Antsaklis P.J. Timing Characteristics of Hierarchical Discrete Event Systems. In Proc. of the 1991 American Control Conference (IEEE Cat. No. 91CH2939-7), p.2917-22 vol. 3.

[19] Ramadge, P.J.G., Wonham, W. M. The Control of Discrete Event Systems. In Proc. of the IEEE, vol 77, no.1 Jan 1989, pp. 81- 97.

[20] A. Sanhez, *Formal Specification and Synthesis of Procedural Controllers for Process Systems*. Lecture Notes in Control and Information Sciences 212, Springer-Verlag London Limited 1996.

[21] Thistle, J.G., Wonham, W.M., Control problems in a temporal logic framework. In Int. J. Control vol. 44 no. 4 pages 943-476.

[22] Ward, P., Mellor, S., *Structured Development for Real-Time Systems*, vol 1-3, Yourdon Press, 1985.

[23] Wilkins, D.E., *Practical Planning. Extending the classical AI planning paradigm.* Morgan Kaufmann Publishers, Inc., 1988.

[24] Brian C. Williams, P. Pandurang Nayak. Immobile Robots AI in the New Millennium. AI Magazine Fall 1996, pp 17-34.

[25] Wonham, W. M., A control theory for discrete event systems. In Advanced Computing Concepts and Techniques in Control Engineering, M.J. Denham, A.J. Laub, editors, pp. 129-169. Springer-Verlag, 1988.

2

CONTROL CONCEPTS AND OPERATIONS WITH PC'S

Ontological control studies the principles, operations, and the architecture involved in the design of a novel type of supervisory controllers called *ontological controllers*. An ontological controller supervises a *programmable controller* (PC) in order to detect violations of ontological assumptions and consequently allows the PC to recover from such violations so that it can achieve a pre-specified goal state.

This chapter introduces first the basic concepts and operations used in control with programmable controllers. The *<condition, control action>* pair is introduced as an appropriate formal construct representing a state of a PC (the so called *controller state*). A goal path is defined as a sequence of controller states (a control sequence) executed by a PC in order to achieve certain pre-specified goal state. Furthermore, two major control operations performed by a PC are presented. These are the so called *synchronization* and the *goal seeking* operation. These two control operations are evoked every time the PC is de-synchronized from the execution of a goal path, that is, it is not able to continue the execution of the goal path and thus, the goal state associated with this goal path may not be achieved.

Secondly, this section explains the notion of a *problematic control situation* (PCS) in the context of a *de-synchronization* from the execution of a goal path and describes some common causes for a de-synchronization. These common causes are first described for relatively simple PCs, namely relay ladder diagrams. Then, they are described for modern, distributed control systems (DCSs). The PCS caused by *violation of ontological assumptions* (VOA) is described by an example and is contrasted with other causes for a PCS. Furthermore, we define four control problems dealing with the description, identification, prevention, and recovery from a PCS. Then, ontological control is defined as the solution to the description, identification, and recovery control problem for a PCS due to violation of ontological assumptions. We conclude the chapter with a number of application areas for ontological control.

2.1 Programmable Controllers

This section introduces informally the basic concepts involved in control applications with a simple type of a PC language, namely, the so called *relay ladder diagram*. The basic concepts introduced are conditions, control actions, control sequences and the interpretation of control sequences. Furthermore, we describe two special operations a PC·uses in the case of a problematic control situation. Such a situation occurs whenever the PC aborts the interpretation of a currently interpreted control sequence. We also describe the causes for aborting a currently interpreted control sequence.

Nowadays PCs control nearly all medium and large scale *industrial processes*. The part of an industrial process which is under the control of a particular PC is called a *plant* and several PCs may work together to control a given plant. A PC is a digital computer equipped with sensors and actuators. The sensors and the actuators are the *process interface* of the PC.

The PC uses sensors to read a number of physical properties of the plant under control, such as temperatures or electrical currents, which are then assigned to variables in the digital computer part of the PC. These variables are called *plant signals*. The set of variables together with their current values are the *process data* of the PC.

Actuators are devices such as relays, coils, motors or breakers. The PC can perform an *on-off* type of control with actuators, or it can change actuator outputs in a continuous fashion.

The digital computer part of a PC uses three main programs:

- *The application program.* The application program is a symbolic representation of the different control sequences which the PC executes. The symbolic representation of a control sequence is realized in terms of the syntax of the particular PC programming language employed. The execution of a control sequence is mapped to a corresponding sequence of plant outputs. The *application program* is specific for each control application. Normally, the application program is designed according to a *control specification* which describes desired sequences of plant outputs.

- *The control data entry program.* The control data entry program, in its simplest form, assists the PC programmer in the interactive design of the application program. The control data entry program verifies that the syntax of the application program is correct. More complex control data entry programs can automate fully or partially the translation of the control specification into an application program. When the control data entry program is active, the controller is said to be in *programming mode*.

- *The interpreter program.* The interpreter program interprets in real time each symbolic component of the application program with the available process data. The interpretation of the sequence of symbols of the application program results in a sequence of actuator outputs which in turn results in a sequence of plant outputs. If the interpretation of the application program and the effects of the actuators on the plant are as intended, the resulting sequence of plant outputs will be according to the control specification. When the interpreter program is active, the controller is said to be in *automatic control mode*.

Historically, applications with PCs expanded during the 1970's and replaced circuits with relays in sequential control applications [13]. The process interface of these earlier PCs remains identical to that of their relay predecessors and consists of boolean signals. A PC that executes its control algorithm by performing operations on boolean signals is called a *programmable logic controller* (PLC).

The most widespread PLC programming language, still in use for small PLCs, is the *relay ladder diagram* (or the *ladder diagram*). Today, standards such as

IEC 1131 ([5], [6], [7]) define the relay ladder diagram syntax and the rules for the interpretation of the ladder diagram programs.

The *syntax* of a relay diagram language consists of a number of graphical symbols which are the building blocks for the control sequences represented in an application program. The *interpretation* of a relay ladder diagram is the operation performed by the *interpreter program* which assigns boolean values to variables associated with each graphical symbol in the application program.

2.1.1 A Relay Ladder Diagram: The Syntax

A *relay ladder diagram* is a symbolic wiring diagram (i.e. a syntactical entity) consisting of a left and of a right power rail symbol, of contact symbols, coil symbols and function symbols. The application program example in **Figure 3** shows these symbols. The control data entry program verifies that the symbols in the application program are correct according to the syntactical rules for the different graphical symbol and associates with each graphical symbol in the application program a name and a variable. However, a graphical symbol has the same name as the variable associated with it. Thus, in the example from **Figure 3**, the contact symbols are those denoted by the variables C_1, C_2, C_3; a coil symbol is the one denoted by the variable $Coil_5$; and a function symbol is denoted by the variable $Timer_4$.

Contact symbols

The contact symbols stand for electrical contacts. A contact symbol can have a designated physical counterpart in the plant belonging to the process interface of the PC. Such a contact symbol represents a physical contact. A physical contact opens or closes depending on the state of its physical counterpart (that is, a state of the plant), this state being represented by a digital plant signal. There may be also contact symbols that have no physical counterparts in the plant and which are associated with coils or functions. In a graphical representation, the association may be done by representing the contact that has no physical counterpart with the same index as its associated coil.

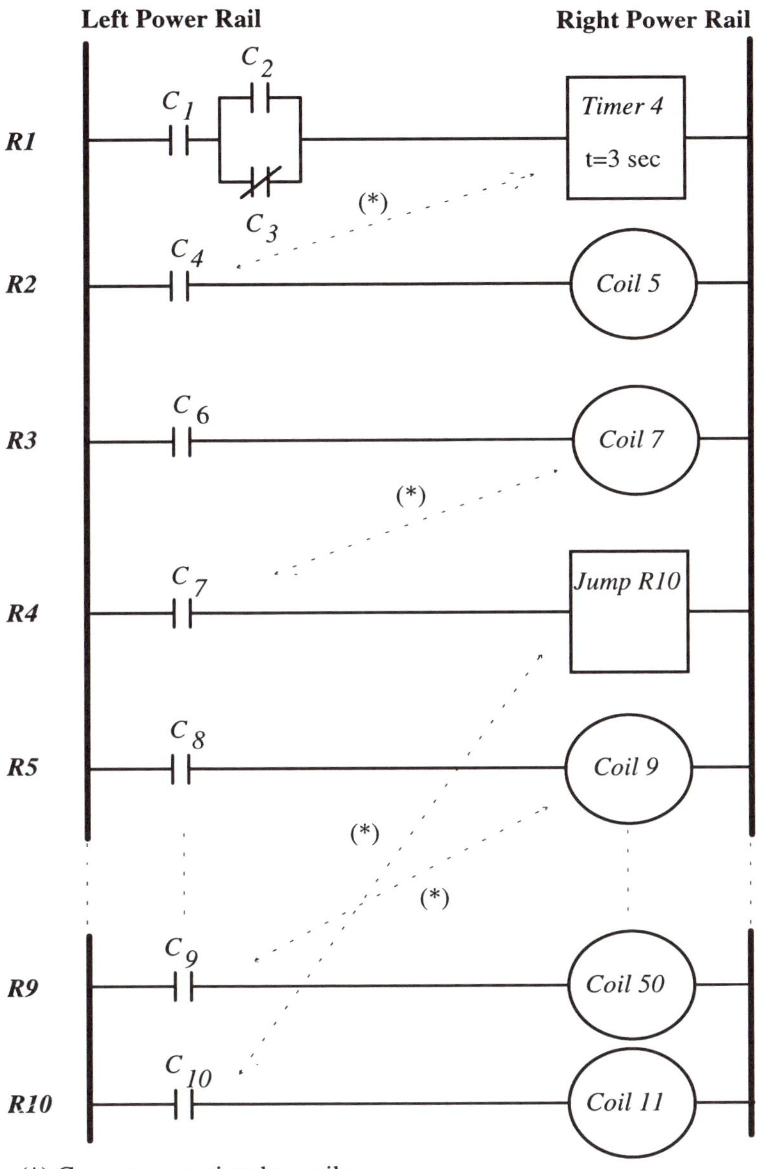

Figure 3 Relay Ladder Diagram.

Thus, these contacts open or close depending on the state of the coil expressed by a discrete signal, or a value of a function (see the section below about *coil symbols* and *function symbols*). In the example from **Figure 3**, the contact symbols denoted as C_4, C_7 and C_{10} have no associated physical contacts since they are associated with functions or coils (*Timer 4, Coil 7*, respectively *Jump R10*). The association is shown with dotted lines in **Figure 3** (the dotted lines are not part of the ladder diagram syntax). The contact symbols can be networked serially and / or on parallel. The syntax requires that at least one contact symbol of the network is connected to the left power rail. A programmer uses the control data entry program to insert contact symbols into the application program, to connect them in a network and to connect the network to the left power rail. The program example in **Figure 3** shows the contact C_1 which is serially connected to the parallely connected contacts C_2 and C_3.

Coil symbols

A coil symbol stands for a physical relay coil. A coil can have a physical counterpart in the plant belonging to the process interface of the PC. There can also be coil symbols with no physical counterparts. The coil symbol is always connected on its left side to a network of contacts and on its right side to a right power rail symbol. Each network of contacts is connected to a unique coil symbol. The programmer uses the *control data entry program* to insert coil symbols into the application program and to connect them to networks of contacts and to the right power rail.

Function symbols

A function symbol stands for a specified mathematical, logical or control function. Each PLC (and PC) has a library of functions. A function symbol is always connected in the following manner: on the left it is connected to a network of contacts and on the right it is connected to a right power rail symbol.

The symbolic representation of a network of contacts connected to a coil or a function symbol, is called a *rung* ([5] pp. 29). A rung stands for an electrical circuit with contacts and an electrical load. We use the notation R_i *(i=1,...,n)* for a rung. In **Figure 3** two examples of rungs are R_1 with the contacts C_1, C_2, C_3 and function $Timer_4$ and the rung R_2 with the contact C_4 and coil $Coil_5$.

Different PC implementations may have other symbols besides those described above. However, the symbols presented above appear in most PCs. Using them, we can now introduce a number of control program constructions which are often encountered in applications.

2.1.2 A Relay Ladder Diagram: The Interpretation

The control data entry program assigns to each contact symbol, coil symbol and function symbol a Boolean variable. The interpreter program assigns Boolean values $(0,1)$ to these variables as follows.

The interpretation of contact symbols

If a contact symbol represents a physical contact, then the interpreter program assigns to the variable associated with the contact symbol the current output of the physical contact. If the contact is of *normal open* type, then it interprets to 0 when the physical contact is open and to 1 when the physical contact is closed. Examples of such contacts in **Figure 3** are C_1 and C_2. A contact may have the negated interpretation (a **NON** function) in which case it may be represented with a crossed line such as C_3 in **Figure 3**. If the contact symbol has no physical counterpart, but it is associated with a coil or a function, then the interpreter program assigns to the variable associated with the contact symbol the value of the variable associated with the coil or the function.

The interpretation of a network of contacts

First, variables associated with each contact symbol are interpreted according to the rules for the interpretation of contact symbols. Then the network is interpreted as follows: the parallel contacts are interpreted using a Boolean OR function and the serial contacts are interpreted by using a Boolean AND function. Each contact may have a NON function, as described above. The result of the interpretation is either 0 or 1. Thus, the interpretation of a network of contacts is identical with the interpretation of a Boolean formula. The Boolean formula representing a network of contacts is called the *condition* of the rung.

The interpretation of a coil or a function symbol

The syntax of the relay ladder diagram constrains each coil or function symbol to a unique rung. The interpretation of a coil or function symbol is the following: the interpreter program interprets the network of contacts of the rung (the *condition* of the rung). The value obtained after the interpretation of the condition is then assigned to the variable associated with the coil or the function of the rung. Therefore, when a network of contacts is interpreted to *1*, the variable associated with the coil of the current rung is interpreted also to *1*. If the PLC has a physical output (e.g., actuator) associated with this coil, then this physical output is activated. Thus, by activating its physical output, the PLC performs a *control action* upon the plant. For example, the activation of a physical output belonging to the process interface of the PLC can start a pump, close an electric breaker or start a heater. Observe here that a control action is performed only if the condition of the rung changes.

The function symbols may have interpretation rules that differ from the interpretation of the coils. For example, the rung with the timer $Timer_4$ from **Figure 3** is interpreted as follows: when the physical contact C_1 is closed and either the physical contact C_2 is closed or the physical contact C_3 is open, the function $Timer_4$ is activated and the non-physical contact C_4 associated with the function $Timer_4$ switches from *0* to *1* after a pre-specified time delay of *3* seconds.

The interpretation of a sequence of rungs

The interpreter program normally interprets continuously the rungs of the application program. When started, it interprets the condition of the first rung of the application program. If the condition of this rung is interpreted to *0*, the interpretation of the same rung continues until the condition is interpreted to *1*. When this condition is finally interpreted to *1*, the interpretation continues with the next rung in the sequence. A sequence is successful when the condition of the last rung, called a goal rung is interpreted to *1*. A sequence that starts with some initial rung and ends with a goal rung is called a *control sequence*. The syntax for how a sequence is defined depends on the particular PC language. Some PC languages have specialized sequential functions while others implement a self-holding coil at each rung in the sequence. Most often only the initial rung has a certain notation (such as a label or index) that can be used as an argument for a jump (see next section for the jump operation). The goal rung is the last rung in a sequence of rungs and thus the contact of the goal rung's coil is not used in any of the rungs of the application program. For the example in **Figure 3**, the rung $R10$ is initial to a sequence while the rung $R9$, placed before $R10$ is a goal rung if the contact of its coil is not used in other rungs.

Interpretation jumps

Normally a PLC has a number of pre-specified control sequences. Each one leads to some goal rung. What particular control sequence is currently under interpretation and consequently, what goal rung will be finally achieved, depends on the current state of the plant. When the plant state changes, the goal rung of the current control sequence may not be possible to achieve anymore, or it may not be the best goal rung to achieve. In such cases, the application program of the PLC can switch from the control sequence already being interpreted to another control sequence which is the proper one for the current plant state. The mechanism for how the switching from one to another control sequence is realized, the reasons for such a switch, and the consequences of this switch are described in the remainder of this section and in the next section.

Each PLC function library has a *jump* function. When the condition associated to a jump function in a rung interprets to *1*, the interpretation continues with the rung specified as an argument of the jump function. Else the interpretation continues with the next rung. In the example from **Figure 3** if the contact C_7 is closed, then the interpretation continues with the rung R_{10}, else with the rung R_5. In the context of this book, the jump function is relevant when used in a *problematic control situation*, i.e., when the interpreter aborts the interpretation of a currently interpreted control sequence.

2.1.3 A Problematic Control Situation

A problematic control situation occurs when the presently interpreted control sequence has to be aborted and a new control sequence is determined. The PC is able to determine a new control sequence by switching over to a special control sequence called the *synchronization sequence* and the consequent use of another special control sequence called the *goal seeking sequence*. In what follows we will describe the causes for a problematic control situation and the use of the synchronization and goal seeking control sequences.

Deadlock

In applications, a plant state may result in some conditions that prevent the interpretation of a control sequence already under interpretation. For example, during an electrical power failure or during a critical alarm for an actuator, there is no reason to interpret a control sequence which depends on the electrical power, or on the correct function of the actuator since its goal rung cannot be achieved. If the interpreter program interprets the control sequence in spite of the power failure or critical alarm indication, the condition of some rung in the control sequence will be interpreted always to *0* and hence the interpretation will be locked to that rung. To avoid this situation, the application program has a rung placed in the beginning of the control sequence. The rung consists of a condition and a *jump* function. The *condition* of this rung, when interpreted to *1*, indicates that the current control sequence is going to be a deadlock. If so, the *jump* function redirects the interpreter program to a new control sequence. For example, this new control sequence can be designated to control in cases of power failures or actuator critical alarms. For example, in **Figure 3** the contact C_7 of the rung R_4 when interpreted to *1* indicates that the interpretation of the rungs starting with R_5 and up to the rung preceding R_{10} results in a deadlock. Therefore, when C_7 interprets to *1*, the new control sequence is the control sequence that starts with the rung R_{10}.

A control sequence which cannot be continued due to a potential deadlock is called a *deadlocked sequence*. A control sequence whose interpretation is aborted, but may be resumed later on, is called a *pending sequence*.

Multiple control sequences

Often several alternative control sequences are available to a deadlocked control sequence. In this case, the application program has a rung containing a jump to a specific control sequence called a *synchronization sequence.* The synchronization sequence determines which is the new control sequence among the available alternatives to a deadlocked control sequence. After the new control sequence is determined, the interpretation continues from the first rung of this sequence. The operation involving a jump to the synchronization sequence and then finding the rung of a new control sequence from where to resume the interpretation, is called *state synchronization operation,* or simply a *synchronization operation.*

When the PC performs a jump to a synchronization sequence it is said to be *de-synchronized from the controlled plant* and we say that a *de-synchronization* occurs. The explanation for why the PC is de-synchronized from the plant is the following. After the jump, the synchronization operation starts interpreting a synchronization sequence rather than a control sequence. Furthermore, the interpretation of the rungs in a synchronization sequence result in finding a new control sequence and not in the execution of control actions that affect the plant. Hence, after a jump and while the synchronization operation is active, the PC cannot affect the plant via the execution of control actions, and thus, it is de-synchronized from the plant. That is why we say that a problematic control situation results in the occurrence of a de-synchronization. Consequently, in the case of de-synchronization, the synchronization operation first performs a jump to a synchronization sequence (thus aborting the currently interpreted control sequence), and then finds a new control sequence by interpreting this synchronization sequence.

For example, let us assume that a control sequence cannot be pursued after a power loss and/or after a critical alarm of an actuator. The application program may have three different alternative control sequences: (*i*) one for power loss only, (*ii*) one for critical alarm only, (*iii*) one for simultaneous power loss and critical alarm. The state synchronization operation may test case (*iii*), then (*i*) and finally (*ii*). As this example shows, the synchronization requires a sequence consisting of three rungs. A single *condition* is not sufficient since it cannot determine which one of the three cases above has occurred.

We would like to stress here that a condition placed within a control sequence is different from a condition placed within a synchronization (or goal seeking) sequence. A control sequence consists of a number of rungs that are to be interpreted in a certain order. When the condition of a rung is interpreted to *1* this corresponds to a control action being executed upon the plant. Each such control action is known to bring about a certain plant output only if the control action is executed under certain preconditions. It is the condition of a rung that is the required precondition for the particular control action corresponding to this rung. In contrast, the condition of a rung with a jump function to a synchronization sequence is the precondition for an entire control sequence, where the interpretation of the rungs in this sequence does not correspond to the execution of control actions upon the plant, but results in finding out a new control sequence. In applications, conditions which are essential for entire synchronization control sequences indicate for example, the occurrence of a power supply failure, critical alarms, etc.

Pending control sequences

A problematic control situation which results in a de-synchronization, takes place when the interpretation of a control sequence is aborted because a condition, reflecting the current plant state, is interpreted to 1 and this particular condition leads eventually to a deadlocked control sequence. However, this plant state may not be permanent. Let us assume that after a time the current plant state is such that a pending control sequence (the aborted control sequence) can be resumed. For example, the electrical power supply may be restored or the critical alarm for an actuator may be over. Due to this, it may be the case that the pending control sequence is more suited or has a higher priority (due to the goal rung it is intended to achieve) than the currently interpreted control sequence, although the currently interpreted control sequence may not be a deadlocked sequence. This problem can be solved with a special purpose control sequence which determines if a pending control sequence has a higher priority compared to the currently interpreted control sequence. If so, a jump is made back to the pending control sequence. Normally there may be several pending control sequences. The operation which re-evaluates the priority of different control sequences, and changes the current control sequence or alternatively allows the continuation of the current control sequence, is called the *goal seeking operation* (GSO). The control sequence that performs the priority evaluation is called a *goal seeking sequence.*

Figure 4 illustrates the synchronization operation and the goal seeking operation. In this figure we use the convention to represent a rung by a circle and two consecutive rungs by circles linked with solid arrows. The dotted arrows stand for additional linked rungs (which are not represented), placed in the same sequence.

Let us assume that the current control sequence is $(R_1, R_2, R_3, R_4, ...)$ and that the rung R_3 is used by the synchronization operation. When the condition of the rung R_3 holds (interpreted to 1), this means that rung R_3 cannot be fully interpreted. Consequently, R_3 executes a jump to rung R_5. The rung R_5 belongs to a synchronization sequence $(R_5, R_6, R_7,...)$. If the condition of R_6 holds, the interpretation continues with the control sequence $(R_8, R_9, R_{10},...)$, else if the condition of R_7 holds the interpretation continues with the control sequence $(R_{11}, R_{12}, R_{13},..)$. At some rung belonging to a later control sequence, for instance R_{17}, the *goal seeking sequence* $(R_{17}, R_{18}, R_{19},...)$ may be executed. At rung R_{18} in this sequence the interpretation switches from the current control sequence back to rung R_3 of the pending control sequence $(R_1, R_2, R_3, R_4,...)$.

Although the goal seeking operation resembles the synchronization operation since both use a jump function, there is still a clear difference between them:

- The synchronization operation is activated when the interpretation of the current control sequence cannot proceed since the continued interpretation would result in a deadlock. The synchronization operation interprets the current state of the plant using plant signals and determines a new control sequence (and implicitly a new goal rung) by taking into account the current plant state. In general, the synchronization operation determines a new control sequence using interpreted plant signals, in a situation (a problematic control situation) when the current control sequence is aborted.

- The goal seeking operation is activated when a choice between several pending control sequences is to be made. The purpose of the goal seeking operation is to find the control sequence with the highest priority among those available such that the interpretation can resume from a rung on this control sequence. In short, the goal seeking operation determines the rung belonging to a pending control sequence which was aborted at a previous point in time. The GSO does not need interpreted plant signals to determine the control sequence with the highest priority, since the priority is given a priori by the control specification and does not change during control.

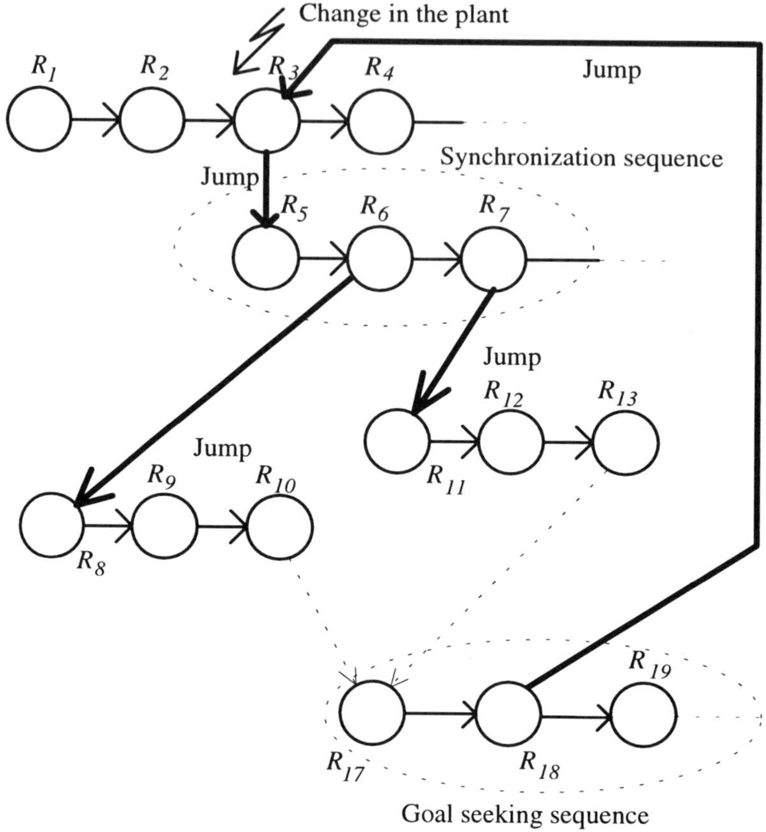

Figure 4 Goal seeking and synchronization with PLCs.

As it is apparent from the description above, we identify three types of sequences in an application program:

- Control sequences (control actions in these sequences affect the plant),

- Synchronization sequences (control actions in these sequences do not affect the plant),

- Goal seeking sequences (control actions in these sequences do not affect the plant),

In Chapter *3* we will describe in detail the relation between the interpretation of control sequences, the synchronization operation and the goal seeking operation. However, even the simple examples from above illustrate that the synchronization and goal seeking operations can be quite complex. The change of the control sequence via a synchronization operation relies on the assumption that the PLC can interpret to *1* each condition of the new control sequence, including the condition of its goal rung. Clearly this assumption may turn out to be wrong. If the synchronization operation of a PLC is not well designed and as a result the interpreter program performs unwarranted jumps away from the currently interpreted control sequence, the synchronization operation is said to be too 'strong'. In this case, the PLC will not reach all the intended goal rungs since the synchronization operation will jump over to another control sequence before the goal rung of the current control sequence is achieved.

Alternatively, if the goal seeking operation is too 'strong' i.e., it prefers pending control sequences with unwarranted high priorities over the currently interpreted control sequence, then the PLC will not be able to pay due attention to current changes from one plant state to another. That is, the PLC will be interpreting outdated pending control sequences and will not be able to respond promptly to current states of the plant.

Finally, if the interpretation operation for a control sequence is too *'strong'* relative to the goal seeking and to the synchronization operations, the PLC may lock-up since the synchronization operation cannot divert the interpretation operation from a control sequence in the case of a problematic control situation. Therefore a properly designed PLC is the one that realizes an optimal relation between the interpretation, the synchronization, and the goal seeking operation. This kind of "tuning" of a PLC is different from traditional control: the former is related to the internal environment of the PLC while the later is oriented towards the external goals to be achieved.

For this book, the relevant aspects of the ladder diagram based PLC are the following:

- A rung can be expressed as a *<condition, control action>* pair. The two parts of such a pair can be identified for each rung of a ladder diagram. The *condition* corresponds to serially and parallel networked contacts on the left side of the rung and is equivalent to a Boolean formula. For the first rung in **Figure 3**, the equivalent Boolean formula is $C_1 \wedge (C_2 \vee \neg C_3)$. The *control* *action*, when interpreted to *1*, corresponds to an activation of an actuator or the execution of a control function among those in the library of the PLC. The control action part is interpreted to *1* (or executed) only when its corresponding condition interprets to *1*. A sequence of *<condition, control action>* pairs defines a control sequence, i.e., a sequence of rungs.

- The order of the *<condition, action>* pairs in a control sequence determines the order in which the PLC performs control actions. Since each condition is associated with a control action, the order amongst conditions determines the order amongst control actions. A control sequence has the objective to reach a goal rung called a *control goal*. A control goal is a special *<condition, control action>* pair, where: *(i)* the condition is the condition of a goal rung and this condition does not appear in any other rung, and *(ii)* the control action is identified with a coil. When a PLC executes the control action of a pair, this is expected to result in a change of the plant state such that the condition part of the consecutive *<condition, control action>* pair will be interpreted to *1*. Due to this property, a *<condition, action>* pair is called a *controller state* and the control goal *<condition, action>* pair is called a *goal state*. A control sequence known to achieve a goal state is called a *goal path*.

- An application program uses two operations which change the goal path currently interpreted by the interpreter program in the case of a problematic control situation. The change of the goal path results in de-synchronization. Both operations use the *jump* function:

(i) The synchronization operation determines a new goal path using interpreted plant signals in a situation when the currently interpreted goal path is aborted (a problematic control situation). The aborted goal path becomes a pending goal path whose interpretation can be resumed at a later point in time.

(ii) The *goal seeking operation* determines a pending goal path whose interpretation is to be resumed and ensures that this goal path has the highest priority amongst the set of pending goal paths.

2.1.4 Summary

In this section we have identified the basic concepts used in control with PCs based on relay ladder diagrams: the controller state, the goal state, the goal path, the goal seeking and the synchronization operations. Furthermore, we described the use of the synchronization and goal seeking operations when a problematic control situation due to deadlocked goal paths results in a de-synchronization. Violation of ontological assumption is yet another cause for a problematic control situation which requires the use of the synchronization and goal seeking operations. However, the significance of this particular cause is apparent only in the context of complex control applications with modern PCs. Furthermore, it is in this type of control application where there is a multiplicity of other causes for a problematic control situation and thus, there is a need to distinguish violation of ontological assumptions from these other causes. Therefore, the following section describes control with modern PCs.

2.2 Modern Programmable Controllers

In this section we reconsider the basic concepts of controller states, goal paths, and the synchronization and goal seeking operations but this time, in the context of control with modern PCs. Furthermore, we re-introduce the notion of a problematic control situation and introduce the notion of a violation of ontological assumptions as one particular cause for a problematic control situation and illustrate it with an example.

The process control industry uses PCs that are a great deal more complex than their 1970's predecessors. The control architecture, the programming language and the programming support of a modern PC makes it the proper tool for both sequential and continuous feedback control applications. Very often, a modern PC is a prototyping tool for embedded control applications, replacing a general-purpose computer with a high level language such as C or Prolog. However, the increase in the size and complexity of the current control applications has lead to a situation when the operator or the commissioning engineer cannot easily determine if a PC is controlling properly or not. A particularly difficult problem is to determine the optimal relation between the operations of interpretation, goal seeking, and synchronization.

A typical control application with modern PCs consists of the components shown in **Figure 5**. Due to its networked architecture, such a system of interconnected PCs is often called a distributed control system (DCS).

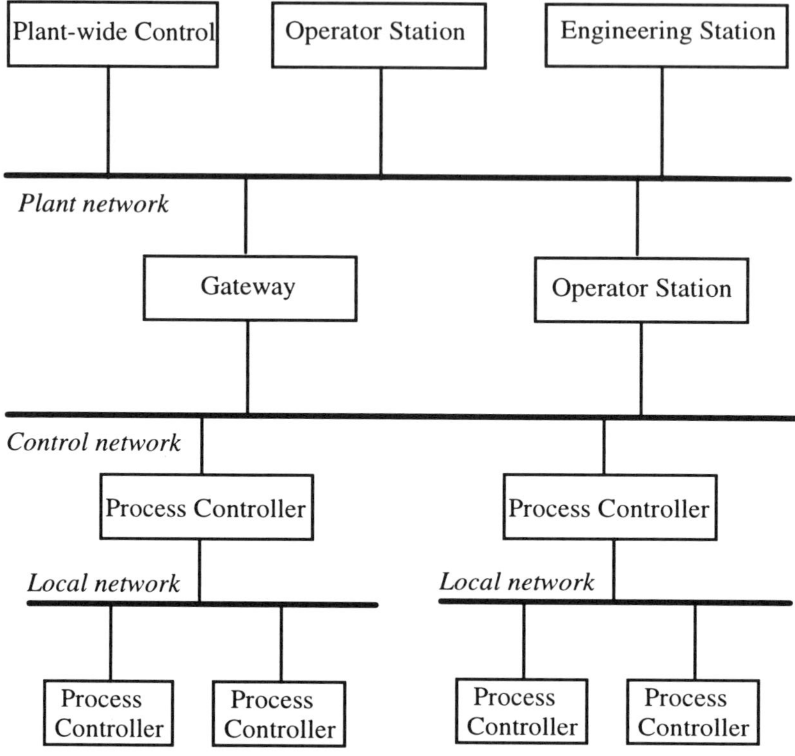

Figure 5 Distributed Control Application.

The components of a DCS are the following:

- *PCs communicating via a hierarchical control network.* Each PC is equipped with sensors and actuators at its process interface with the controlled plant (not shown in **Figure 5**). Typically a medium sized PC may have several thousand I/O signals connected to actuators and sensors. A modern PC can integrate both sequential and continuous feedback control algorithms. The process signals are both analog and digital.

- *Operator stations.* An operator can give commands and visualize the current state of the PC and/or the plant by means of graphically represented models, time-series depicting trends of PC and plant parameters (logs), various processed data statistics and alarms.

- *High speed communication networks.* High speed communication networks connect PCs and operator stations. A PC may communicate to other units of a superior system, to peer units and to subordinated units. The information exchanged between PCs through the network can be:

1. *Plant values*, e.g. temperature, pressure, electrical currents.

2. *The operating status or mode of the PC*, e.g. data showing if a PC is operational or not; if it is in *manual control* mode or in *auto control* mode; if it is in a particular startup mode *(cold, hot, warm)*, or if it is switching from one to another mode of operation.

3. *Operator commands.* There are several levels of operator commands such as:

   - Commands to change the value of an I/O (I/O forcing).

   - Commands to change the current state of an application program (state forcing).

   - Commands to change the operational mode of a PC (restart a PC in different modes (cold, hot, warm), or to set a PC in manual or auto mode).

   - Commands to change program parameters, or to change fully or partly the currently executed PC application program.

4. *Error messages*, e.g. internal failures or timing errors such as violations of the real-time requirements.

5. *Synchronization signals* for parallely executing application programs *(tasks)* in PCs.

6. *Synthesized information*, e.g. data from process loggers and statistical data.

Since several PCs control the same plant, they normally share some data via a network. For example, one PC may need the operational status of other PCs. The hierarchical configuration with networks means that one PC in the network has as inputs not only signals from the controlled plant but also data from the control operation of another PC. Reciprocally, a PC may act not only directly upon the controlled plant but also upon a lower level PC. Therefore the actuator of a PC may be another PC at a lower level in the control hierarchy. Equally, the process input of a PC may be complex information from a superior PC.

The programming languages for PCs are adapted for a hierarchical control architecture. Although there are many programming languages available, some of their features are common. Among them, those relevant for this book are:

- The programming languages for PCs accept both sequential control algorithms and continuous feedback control algorithms.

- A PC language allows a limited degree of control on parallely executed application programs ('limited' as compared to multitasking operating systems). Parallel application programs may exist either in the same PC (an application program is then called a *task*) or in physically separated PCs, communicating via the network. An application program can initialize, start and stop another application program in the same PC or another PC. Moreover application programs executing in parallel can synchronize their execution (i.e. one application program waits until another application program is in a predefined state). This means that one control action in an application program can be a goal path in another application program.

- A PC programming language allows implementation for time-critical control. For example, it can enforce that a particular application program is executed completely within a predefined time interval, or it can ensure that transitions in a goal path occur within specified time constraints. If the timing requirements cannot be met, the controller can perform actions, such as to stop or increase the time intervals for low priority tasks.

An example of a modern PC programming language is the so called *function block diagram* (FBD). The basic elements of a FBD language are defined in standards [5], [6], [7]. The language consists of graphically represented function blocks (or elements) which have inputs and outputs. There are available elements for different types of control such as time critical control, continuous control, sequential control, task control, as well as elements performing arithmetic and Boolean operations or serving as function generators. The PC control program consists of a set interconnected elements such that the output of one element is the input of some other element. For example, the rung R_1 in **Figure 3**, with the logical function $C_1 \wedge (C_2 \vee \neg C_3)$ and timer function $Timer_1$ can be represented in an FBD language (using ISO logical gate symbols) as shown in **Figure 6**.

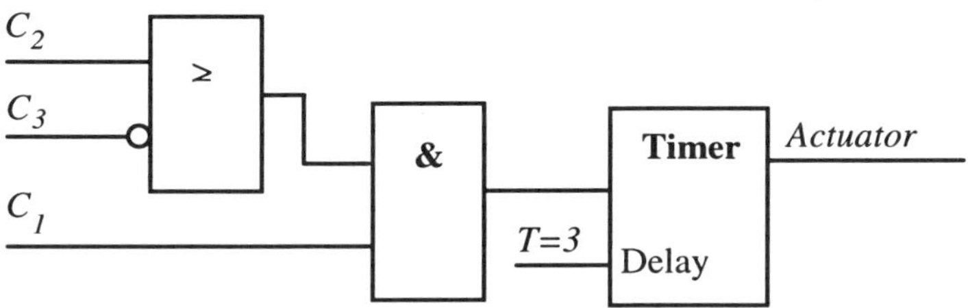

Figure 6 Logical and Timer Function with FBD.

The plant inputs and outputs have also corresponding elements. **Figure 7** shows a PI control function using function elements from [1] pp. 93.

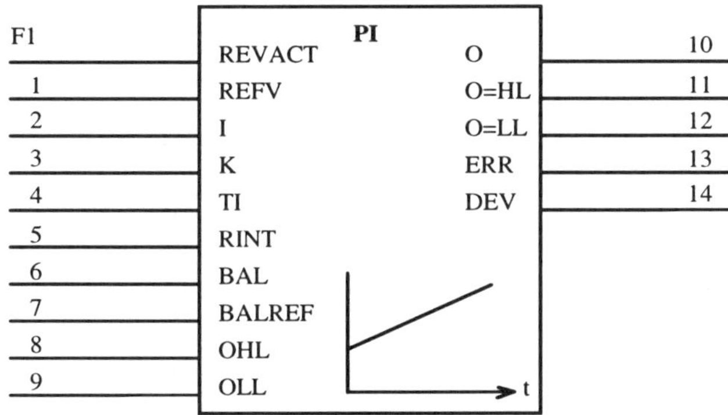

Figure 7 PI Controller.

Table 1

No	Name	Type	Application
F1	REVACT	Boolean input	Reverse action. REVACT=1 gives inverted control action
1	REFV	real input	Reference input value
2	I	real input	Input of the actual value
3	K	real input	Gain setting
4	TI	time real	Time integral constant
5	RINT	Boolean input	Reset integrator
6	BAL	Boolean input	Input for activation of tracking
7	BALREF	Real input	Input for tracking reference value
8	OHL	Real input	Output high limit setpoint
9	OLL	Real input	Output low limit setpoint
10	O	Real output	Output control signal
11	O=HL	Boolean output	Output set to 1 if O reaches HL
12	O=LL	Boolean output	Output set to 1 if O reaches LL
13	ERR	Boolean output	Output set to 1 if OHL<OLL
14	DEV	Real output	Output for control error

The inputs and outputs of the PI regulator are shown in Table 1. Observe in this table the existence of some built-in functions intended to avoid some common causes for a problematic control situation. These functions are modeled by means of Boolean signals. One example is the BAL (balance) Boolean input. This input activates a tracking of the BALREF (balance reference) signal on the output O. The BAL input should be active when the PI control is not active. This ensures a bumpless transfer on O when the control is resumed. The RINT reset integrator signal can be used to reset an integral windup. Other Boolean safety functions are the high limit HL indication and the low limit indication LL for the control output.

An important function block for an FBD language is the *sequential element.* A sequential process is implemented using a number of sequential elements. The element of one step in the control sequence specifies the activation condition for that element (i.e. the required condition for the execution to proceed with the next element in the control sequence) and the timing constraints for that particular step.

Each modern PC language accepts both continuous control loops and sequential control in terms of control elements. An FBD language provides a balanced programming environment for both sequential and continuous controls.

Some PC languages are better suited for continuous control loops while others are better suited for sequential control. For example *Grafcet*, a language based on Petri nets, is well suited for sequential and parallel control tasks (resource sharing, scheduling, etc.) [8].

Although the syntax of the FBD differs from that of a ladder diagram and even though typical FBD applications are mixed (continuous feedback control and sequential control), the FBD goal path is the same as for a ladder diagram and consists of a sequence of *<condition, control action>* pairs. These pairs meet an intuitive control approach [4]: a control action is executed whenever predefined conditions are satisfied such that the result of the control action will take place as expected. More specifically, close to the ladder diagram programming style, the FBD language uses for the condition part of a *<condition, control action>* pair Boolean functions (*AND, OR, XOR*, etc.), which are equivalent to the network of contacts for a ladder diagram. What is new for the condition in a modern PC language compared to the ladder diagram language is the different nature of the arguments for the Boolean functions and the type of control actions the PC can execute. Some examples of conditions and control actions which are non trivial for earlier PCs, but which are normally available in modern PCs are:

- A PC can stop low priority control tasks when the control processing unit (CPU) of the PC is loaded above a predefined limit. This type of control action preserves the real-time requirements for the critical tasks of the controller. The condition in this case is a relation using a measured value of the CPU load of the controller. The control action determines which task has low priority and then stops or delays it.

- Many PCs allow that input/output boards or application programs are changed while the rest of the application program and/or hardware continues to control (so called *on-line pluggable equipment* and *hot program changes*). A condition can test the occurrence of such a change and the control action can ensure that the PC acts safely during the change.

- A condition can test if after a failure the PC can be cold-[1], hot-[2] or warm-restarted[3]. The condition in this case is a function of factors such as the time

---

[1] Cold-restart: restart of the PLC and its application program after all dynamic data (variables such as I/O images, internal registers, timers, counters, etc., and program contexts) are reset to predetermined state.

[2] Hot-restart: Restart after power failure which occurs within the process dependent maximum time allowed for the PC system to recover as if there has

interval during which the PC was stopped, the integrity of its internal data, the difference between the data from the controlled plant and the internal data of the PC, etc. The control action realizes one of the above types of restarting the PC.

- Tests can determine a condition indicating whether two parallel application programs (in the same PC or distinct networked PCs) co-operate regarding some control goal or not. If this is not the case, the state of an application program can be modified such that the two programs will begin to act in co-operative fashion. This modification is performed by an appropriate control action.

For a better understanding of the type of control encountered in control applications with modern PCs, we give a short description of some typical control applications.

Example 1. Flexible manufacturing system (FMS)

This example describes a typical sequential control application. A *flexible manufacturing system* (FMS) consists of a number of robots that assemble or manufacture a product in co-operation with each other. The group of co-operating robots is called a *factory automation cell*. Within a factory automation cell, a work piece is moved from one robot to another one and each robot performs a specific operation on it, e.g., assembly, painting, etc. Each robot has its own PC controller for grasp and arm control as well as for a sequential control of the sequence of operations which the robot can perform on the work piece. The robots in the cell are reconfigurable (i.e., for the type of arms and tools they use) and programmable (for the type of products they assemble). A main PC, called the *cell controller*, is responsible for loading the right program into each robot, starting the robots, and coordinating their operation.

One problem encountered in the design of the cell controller is the following: how to ensure that the processed parts have a transfer rate amongst robots which avoids queues or unnecessary long delays? Since each robot has its own PC it takes a variable time interval, within some specified limits, until it can complete

---

been no power failure. All I/O information and other dynamic data as well as the application program context are restored or unchanged.

[3] Warm restart: restart after a power failure with a user programmed predetermined set of dynamic data and a system predetermined application program context.

an operation or a sequence of operations on the work piece. Therefore, the cell controller has to coordinate the transfer of parts amongst the robots and the start time for the sequence of operations to be performed by each of the robots.

Another problem arises when one type of a workpiece is replaced with a new type of a workpiece. In this case there is the so called transition period during which both types of work pieces are in the work cell. In this context, the cell controller has to start unloading the control program for the processing of the old workpiece and start loading the one for the new type of a workpiece without stopping the cell. Let us assume that the old workpiece is coming back to a robot for further processing after some other robots have already finished their operations on it (i.e., one and the same robot executes two or more different types of operations on the same workpiece at different points in time). Assume also that these other robots have started operating on the new work piece. While in this transition period when both new and old types of work pieces are in the cell, the cell controller has to load the control program realizing an operation on the new type of workpiece and then load back the program for the right operation on the old type of a workpiece.

Another non-trivial sequential control problem is how to restart the cell after a failure.

For the FMS application, the sequential program of each robot and the continuous control loops for grasp and arm control is an issue for the design of the PC for that particular robot. At the level of the cell controller, a particular robot is just an actuator executing a well-defined sequence of control actions subject to certain timing constraints.

Example 2. Diesel generator control

A power plant uses several diesel engines, each one with its own electrical power generator. Each diesel engine has its own PC for the following types of control:

- To execute the start sequence of the diesel engine, i.e. to start the engine with an air compressor, then to supply the engine with diesel oil and finally with heavy oil.

- To synchronize the frequency and phase of the generator to the common electrical power line.

- To load electrically the generator.

- To execute a stop sequence consisting of electrical unloading, electrical disconnection and engine stop.

The start sequence uses a continuous speed regulator which ensures that the speed trajectory follows a predefined pattern for the given engine type. Moreover the frequency synchronizer uses two feedback loop regulators to adjust the frequency and the phase of the generator to the common electrical power line.

A power plant has a main PC. The main PC adjusts the number of diesel engines delivering electrical power considering the global electrical power requirements. For the main PC, the PC of a particular generator plays the role of an actuator. The sequential program of the main PC executes a control sequence for starting and stopping the diesel engines by taking into account the minimum and the maximum times an engine should be run, the preferred priority of starting the engines, and the minimum number of engines that should run simultaneously. Some relevant control tasks for the main PC are the following.

- *A timing task*

Due to physical and thermodynamical constraints, a generator needs certain time to start. If there is an increase in the electrical load demand lasting for a short period of time, it may happen that a generator reaches the power delivery mode at a time when the load demand is again low. This situation is not desirable and should be avoided.

- *A safety task*

 The power delivery should be uninterrupted. Therefore the main PC should be robust in the case of events such as unexpected mode changes in the operation of the diesel engine, diesel engines that do not start, etc. When the main PC, or the diesel engine PC are switching operation mode (i.e. between *manual mode* and *auto mode*) or are restarted after failures, they are expected to adapt immediately to the current plant situation without creating operational risks.

As the examples above suggest, the sequential part of a PC control program is often much larger than the feedback continuous control part. The control action of a <condition, control action> pair in a control sequence may perform a very broad range of tasks. A control action can be a simple *close* command for a breaker, it can be a command to start a feedback regulator, or it can be a command to start another PC. For a modern PC application, the interpretation of the current goal path can be interrupted by numerous, diverse in nature, and many times are hard to anticipate causes. Such causes are: failures in the plant or the actuators, communication errors, resource limitations, sudden changes of operating modes, unexpected start-up of inactive controllers, on-line hardware changes, hot application program changes, etc. Therefore, for a PC controlling a complex control application, the design of the goal seeking and synchronization operations which can cope with these causes is normally a more complex task than the design of the goal paths which include control actions executing feedback loops and sequential control. In other words, the *complexity* associated with a control system does not mean solely a large number of I/O points or a large number and / or a variety of control algorithms. The complexity is a result of the intricated time evolution of the goal paths. The time evolution is complex due to causes that interrupt the interpretation of the currently pursued goal path and thus, a switch over to new goal paths takes place via the synchronization operation, while due to the goal seeking operation pending goal paths can be reactivated.

In short, the features of a modern PC that are relevant for ontological control are the following:

- *State representation.* An application program can be represented as set of sequences of *<condition, control action>* pairs. A *<condition, control action>* pair is called a *controller state.* A sequence of *<condition, control action>* pairs is called a control sequence. The last *<condition, control action>* pair in a control sequence which condition part represents a particular control goal and does not appear in other control sequences is called a *goal state.* As for the ladder diagrams, a control sequence with a *goal state* is called a *goal path.*

- *Parallel execution.* Each PC can execute a number of parallel goal paths (each goal path is called a *task).* A number of PCs may work in parallel.

- *Hierarchical conditions.* The condition of a *<condition, control action>* pair can represent a condition on state of the plant under control, or a condition on a state of the software or the hardware component of a PC. As already shown, a condition test can be performed for a plant state, a state of a parallel program in the same PC, a state of the PC as such, a state of a peer, superior or subordinated PC.

- *Hierarchical control actions.* The control action of a *<condition, action>* pair may have direct effect on the plant by means of actuators. Alternatively the control action consists on messages passed among tasks in the same PC or between a task in the same PC and a task in some other PC.

- *General purpose goal seeking and synchronization operations.* The number and type of causes for a problematic control situation is very large. The occurrence of a cause has the effect that the interpretation of a current goal path cannot be continued and a de-synchronization occurs. Then the synchronization operation, taking into account the current state of the plant, first finds a new goal path whose goal state can be achieved, and then switches the interpretation over to this new goal path. The old goal path becomes a pending goal path whose interpretation may be resumed later. There may be a large number of pending goal paths and this complicates the choice of a particular pending goal path by the goal seeking operation. A modern PC can use several methods to switch the interpretation from one to another goal path. Mostly used are state forcing for Grafcet, controlling task execution or performing jumps among control sequence elements for a function block language. Although the principle for switching the goal path is the same for modern PCs as for ladder diagram-PCs, the number and the nature of causes that may occur and the parallel execution of programs require that a PC control program has a general purpose synchronization and a goal seeking operations.

- *Controlling under assumptions.* A particular PC, due to real time performance and resource limitations, cannot have all the states of the plant and of the other networked PCs represented explicitly in its application program. The plant states read by another PC may not be accessible, or if they are, the information about them may arrive too late over the network. Moreover one PC has memory limitations and it cannot store the state set of all the other PCs in the network. Therefore one PC cannot act based on the state set of the overall control system.

Thus, for each control action the PC executes, there exist normally a number of assumptions about the states of the other parts of the overall control system or of the plant that are not explicit in the condition part of the *<condition, control action>* pairs. However, if these implicit assumptions do not hold when the corresponding control action is executed, the expected effect of the control action upon the plant will not be realized. These implicit assumptions are called *ontological assumptions.* When the ontological assumptions are not fulfilled, the PC is said to act under a violation of ontological assumptions.

- *Violation of ontological assumptions.* In the case of a violation of ontological assumptions, as described above, the PC will not realize the expected effect of its control action upon the plant. Thus, the condition part of the controller state consecutive to the current controller state cannot be interpreted to 1 after the execution of the control action from the previous controller state. This situation however resembles the case described earlier when the occurrence of a cause such as a potential deadlock renders the interpretation of a goal path impossible. Therefore, it is the state synchronization operation which will be activated also in the case of violation of ontological assumptions. However, violations of ontological assumptions are not amongst the causes for which the application program provides alternative goal paths (via the jump function or some other means). Therefore, if the synchronization operation does find a new goal path, the goal state of the goal path whose interpretation was aborted due to a violation of ontological assumptions cannot be achieved anymore. This is due to the fact, that this goal path is not considered a pending goal path because the jump from it to the new goal path was not specified a priori in the application program.

2.2.2 An Example With Ontological Assumptions

The following example illustrates informally the concept of ontological assumptions. More detailed examples are presented further on in the book.

An electrical power generator unit consists of a diesel engine and an electrical generator linked by a clutch (**Figure 8**).

A speed controller is employed to meet the operational requirements for the diesel engine and for the generator as follows. The output from the speed controller is the command signal to an actuator which changes the speed of the diesel engine.

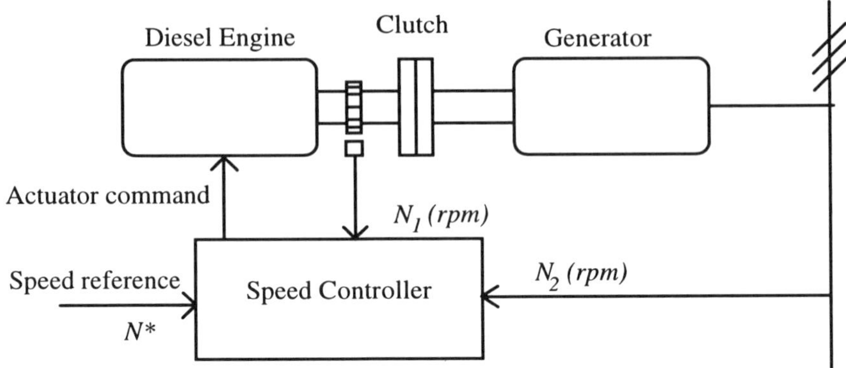

Figure 8 Power Generator Control.

The speed controller determines the engine speed using two physical principles: when the diesel engine has low speed, a proximity sensor reads the rotation speed from a teeth wheel placed on the engine shaft. The corresponding value is N_1 [rpm]. Due to the reduced number of teeth, this measurement gives a speed value with large error limits. When the diesel engine rotates with a speed larger than a specified limit, N_{lim}, the speed is measured more precisely using the frequency of the generated electrical current according to the formula $N = 2 \times 60 \times f \div p$ (p is the number of magnetical poles of the electrical generator and f is the frequency). This value for the speed is N_2 [rpm]. The speed controller uses the values N_1 and N_2 to control the speed in two modes defined as follows:

Control mode 1.

At start-up, the speed controller accelerates the diesel engine according to a predefined start-up trajectory, using the feedback speed value N_1. The goal state for this mode is *a low error deviation between the predefined speed trajectory and the actual engine speed.* This mode is used for speed values in the range $0, ..., N_{lim}$.

Control mode 2.

When the diesel engine reaches the speed value N_{lim}, the controller enters the speed control mode with N_2 as speed feedback value. The goal state for this mode is *a constant engine speed N^*, specified by the reference speed input.*

For this application, one ontological assumption is implicit in the formula $N = 2 \times 60 \times f \div p$ which computes the speed of the engine by using the frequency. This ontological assumption is the following: *both the diesel engine axle and the electrical generator axle have the same rotation speed, since they are linked.* If this assumption does not hold and the speed of the two axles is different, then the formula above does not represent correctly the relation between the rotation speed and the frequency.

Let us assume that there is a violation of this ontological assumption. For example, the clutch between the two units is faulty and it does not hold firmly the two axles together. Therefore, the speed of the diesel engine is always higher than the speed of the generator: $N_1 > N_2$. Let us follow now a control sequence for the diesel engine.

The controller starts the diesel engine in *mode 1* using the predefined speed trajectory. When the controller reads the current value N_1 such that $N_1 = N_{lim}$, it switches to *mode 2*. However, at this moment, the speed controller is reading a speed $N_2 < N_1 = N_{lim}$ and consequently it switches back to the mode *1*. In *mode 1*, the controller reads again that $N_1 = N_{lim}$ and consequently, it switches back to *mode 2*. The mode switch described above is repeated continuously and the controller cannot achieve any of the goal states specified for each of the two modes, but will instead produce continuously a speed around the N_{lim} value.

From this example it can be seen that there is a distinction between the case when a control action does not produce the expected result due to violations of ontological assumptions and the case when the action does not occur due to say wrong physical laws or control algorithms. In the example above, the formula that computes the speed using the frequency value is correct. Therefore when the control action brings the engine speed to the value N_{lim}, the value N_2 is expected to be equal to N_1. This expectation fails. Yet the physical laws underlying formulas are the right ones for this particular control application. However, at a particular time point the assumptions under which the formulas are used, are not satisfied anymore and therefore the controller cannot achieve the goal states.

In this book we consider problematic control situations due to a variety of causes, including violations of ontological assumptions. Since a problematic control situation results in a de-synchronization, the causes for a problematic control situation are also the causes for a de-synchronization. In general, the concept of a problematic control situation does not have a precise meaning since it is relative to the current control technology, and practice. However, for the particular case of control applications using PCs, the notion of a *problematic control situation* is defined as follows:

(1) At a certain time point, a PC aborts the interpretation of a goal path due to a cause, which is not a priori specified in the application program of the PC.

(2) The application program of the PC has no a priori specified means of selecting a new goal path after the occurrence of the unspecified cause.

For example, let us assume that a PC is interpreting a goal path. Meanwhile, the following event occurs: the PC loses for a short time its electrical power supply, so that some internal data used by the interpreter program is retained while some other data is lost. When the electrical power is restored, the interpretation cannot be resumed directly since the condition in the current controller state may not correspond to the actual plant state due to the data loss. The interpretation may be continued, for example, only after the dynamical and I/O data, (such as the values of integrators and the plant output), are updated. If the PC has an application-independent method that can restore the controller data after power losses, then there is no problematic control situation. However, if such method is not available, then there does exist a problematic control situation since the PC has indeed the specified process-related control sequences but none of them is appropriate to take over the control after a power loss of the PC unit.

One objective of this book is to describe the principles that can be used to distinguish a problematic control situation due to violations of ontological assumptions from a problematic control situation due to other causes. To this end, the following section presents a classification of the causes for a problematic control situation.

2.3 Causes For A Problematic Control Situation

In this section we provide a classification for the causes of a problematic control situation. We consider these causes for both continuous feedback control and sequential control. Furthermore, we define four major control problems related to control in the case of a problematic control situation and describe some of the existing solutions for these problems in continuous feedback and sequential control. We conclude by specifying the causes for a problematic control situation which are the subject of this book as well as the control problems related to these causes and dealt with in the book.

Although traditional control problems related to stability, non-linearity or lack of precise analytical models still do exist, there are new problems emerging, which are specific for large industrial process control applications using PCs. These problems increase with the complexity and the size of the application. The experience shows that for typical large industrial control applications such as those encountered in electrical power generation, the pulp and paper industry, integrated marine control, manufacturing, etc., the costs for solving the sequential control problems during the design, commissioning and test of a complex PC program is well beyond the costs related to continuous feedback control problems. That these new types of problems do exist is recognized in a number of papers [2], [3], although there is insufficient information available that would indicate a definition or a classification of these problems.

In what follows we will concern ourselves with four control problems which arise in the context of a problematic control situation. These of course are only some of the control problems encountered in large industrial control applications and our discussion of these four control problems is based on how they are actually approached in practice.

- **The description control problem:** Is there available a (formal) description of the causes for a problematic control situation?

- **The prevention control problem:** What is the principle and method that can ensure that the cause for a problematic control situation does not occur?

- **The identification control problem:** When the cause for a problematic control situation cannot be avoided and as a result it occurs, how can it be recognized that the problematic control situation is due to exactly this cause rather than to some other cause?

- **The recovery control problem:** How can the control be resumed in a problematic control situation due to a known cause? The control situation at the time instance of the recovery imposes constraints, since a problematic control situation may be recognized only after it has actually occurred.

In what follows we will describe the causes for a problematic control situation and some of the existing solutions for the control problems from above.

2.3.1 Causes In The Case Of Continuous Feedback Control

In what follows we will describe some of the major causes for a problematic control situation in the context of continuous feedback control as well as the existing solutions for some of the four control problems from the previous section.

(1) Plant-related causes

Non-linearity, varying environment conditions, long time delays (dead-times) and internal process couplings are the most often encountered causes for a problematic control situation ([1] pp. 25).

- *Non-linear plants.* The formal representation for non-linear plants is in terms of non-linear differential equations constituting the model of the open loop system. The controller design technique is traditionally based on a linearization of the open loop system, a control law depending on the type of linearization, and a consequent stability analysis of the closed loop system around a given operating point, or a trajectory. The problematic control situation can be identified as unstable behavior and its cause is a linearization which is not a satisfactory approximation of the original non-linear model. Often a global stability analysis can be done using the Liapunov function.

- *Varying environmental conditions.* A problematic control situation due to not taking into account the time varying environmental conditions can be prevented by employing non-linear differential equations with time-varying coefficients. Another y to prevent the problematic control situation is the use of *gain-scheduling control* ([13] pp. 27, 303).

- *Time delays.* A problematic control situation due to time delays can be prevented by incorporating these time delays in the model. Depending on the nature of the time delay, solutions for the prevention control problem in the case of causes such as time delays are the *Smith-predictor* (a model-based control) ([13] pp. 25, 301) and *feed forward control* ([13] pp. 25).

An often used solution to the recovery control problem in feedback control is controller *auto-tuning* and *adaptive control* [23].

(2) Signal processing-related causes.

A signal processing cause for a problematic control situation is the existence of *alias frequencies* due to low sampling frequency of the input continuous signals ([13] pp. 130). A solution for the description control problem is analytical, and expressed by the Nyquist relation between the sample frequency and the highest frequency component of the sampled signal. The solution to the prevention control problem consists of selecting a sampling frequency according to the Nyquist sampling theorem and using anti-alias filters, such as the Butterworth filter.

(3) Implementation and computational related causes.

Most often met implementation and computational causes for a problematic control situation are the bumpless transfer, integral windup and delays due to the time required for numerical computations. The bumpless transfer and integral windup causes can be described in a qualitative fashion, while delays due to computations can be modeled. The solution for the prevention control problem for a problematic situation due to bumpless transfer is, for instance a continuous reference follower. The integral windup can be prevented by *conditional integration* ([13] pp. 287).

2.3.2 Sequential Control

In what follows we will present the causes for a problematic control situation in the context of sequential control with modern PCs as well as some of the existing solutions to the four control problems related to a problematic control situation.

(1) Reachability-related causes.

There are several established representations for sequential control programs such as discrete event systems (DES) [21], [17], Petri nets [14], and finite automata [4]. The causes for a problematic control situation can be related to the impossibility of fully traversing specified control sequences. Particular causes are thus the non-reachability of states and deadlock. Results are available for the solution of the identification control problem and the prevention control problem in the theory of Petri nets [14] and discrete event systems [17]. An approach which allows the formulation of a number of prevention principles is the automatic synthesis of sequential control schemes [9]. Then the prevention control problem can be solved with an adequately specified set of goal paths and a set of initial controller states. There are few results available about solutions to the recovery control problem in the case of an aborted goal path. In the Ramadge-Wonham [16] paradigm of supervisory control, a problematic control situation is modeled by the occurrence of a so called *"uncontrolled event"*. A result on the recovery control problem for hierarchical DES is reported in [20].

Although many problems of this class have well-defined solutions, these results are not yet common in applications. For example, the formal approaches to the specification and verification of requirements as presented in [15] are hardly used. The recovery control problem would benefit from systematic methods of requirements revision as reported in [19]. Moreover, there are a number of problems often met in applications which are still subject for ongoing research. For example, a problem, apparent from the examples presented previously, is the so called hierarchical coordination. Results for hierarchical coordination are reported for delay-free DES systems [22].

(2) Timing related causes.

A sequential control (as used in PCs) may not fulfill pre-specified timing constraints and this can cause a problematic control situation. The four control problems related to a problematic control situation due to a timing cause cannot be solved solely with sequences of *<condition, control action>* pairs. Even if the PC control program specifies a number of possible goal paths with reachable and live *<condition, control action>* pairs, this is not enough to ensure that the goal path can indeed be completely interpreted. There may exist additional timing requirements. For example, the information about controlled objects may be out of date or an actuator may not act within a pre-specified time interval.

Solving the control problems due to timing causes requires additional information. Normally there is a specification for the allowed time intervals during which the PC must achieve a goal state, and constraints on the time intervals for the transition between two consecutive *<condition, action>* pairs. When the time intervals are not expressed in absolute time units but specified by temporal logic operators (i.e. 'O' - the 'next' operator, '□' - the 'henceforth' operator, and '$\mathcal{U}$', - the 'until' operator), the relation between time intervals can be extracted from the control specification and a proof for the correctness of the application program of the PC with respect to the temporal relations between the time intervals can be done in temporal logic ([18], [12], [10], [11]).

A problematic control situation due to timing leads to aborting the interpretation of a current goal path. A solution to the recovery control problem related to a problematic control situation due to the above cause is described in Section 5.3.4 and this solution is given in terms of the *state synchronization operation*.

Another solution to the four control problems related to a problematic control situation due to timing, which is still a subject for ongoing research, is the so called *duration calculus*. A control problem description with duration calculus uses time represented via the accumulated presence of a state. This representation is very close to the time representation used in PCs. Specifications for real-time systems with formulas in duration calculus can describe critical durations, progress from state to state and stability of a state [15].

(3) Ill-represented state related causes.

An *ill-represented state* is a state for which the condition part of a *<condition,control action>* pair does not reflect the actual plant state at a certain point in time. In other words, the PC interprets the condition part of a state to *1* at certain point in time, but the actual plant state is different from the one represented by the condition part .

We distinguish two cases when a condition does not correspond to the actual plant state:

- *(i)* Some subformula or a relation that appears in the condition does not reflect properly the physical laws, or other type of knowledge used for the specification of that particular condition. For example, a *'sine'* function is used instead of the '*cosine*' function, or a '*less than*' relation is used instead of the right '*higher than*' relation.

- *(ii)* The condition is expressed with a correct formula, but the interpreter program interprets the condition with old values for plant signals, which do not reflect the current plant state. For example, the integrated value of error may still be there after a PC has been switched from manual mode to automatic control mode. In general, the plant signals which are estimates (obtained via an observer) of internal plant variables, when not updated right, may lead to ill-represented states.

In the book we consider only ill-represented states according to case *(ii)*. The states which are ill-represented according to case *(i)* can be detected by a direct verification.

An ill-represented state leads to the following. Let us assume two consecutive controller states on the same goal path, $<condition_1, action_1>$ and $<condition_2, action_2>$ and let $condition_1$ be ill-represented. Since these controller states are consecutive, after the interpreter program interprets $condition_1$ to 1 at time t_1 and the control action $action_1$, is executed, it is expected that $condition_2$ will hold at some time $t_2 > t_1$. However, since $condition_1$ does not represent the actual plant state, the control action $action_1$ will not produce the expected effect upon the plant and $condition_2$ of the second state will not be interpreted to 1 as expected. As a result, the interpretation of the current goal path is interrupted since a condition on this path cannot be interpreted to 1.

An ill-represented state cannot be identified from a syntactical description of goal paths consisting of $<condition, control\ action>$ pairs, nor can it be discovered by using extra timing information. A solution to the identification problem in the case of an ill-represented state is to use a set of consecutive $<condition, control\ action>$ pairs obtained from the real execution of system (called *traces* or *histories*). Then these traces have to be analyzed by comparing them to the traces that can be obtained from the application program or the control specification for the application program. For industrial real-time systems, traces are normally not available in a data format that is usable with PC application programs. However, the difficulty is not in the design a particular control architecture that can make traces available, but is due to the lack of general solutions for the identification and recovery control problems for PCs when the problematic control situation is due to ill represented states.

(4) Violation of ontological assumptions related causes.

A controller state with a violation of ontological assumptions is a controller state whose condition, after a VOA has occurred, does not correspond anymore to any physically possible plant state and therefore it cannot be interpreted to 1. Let us assume two states on the same goal path, denoted S_1 and S_2 such that $S_1=\langle condition_1,\ action_1\rangle$ and $S_2=\langle condition_2,\ action_2\rangle$ and let S_2 be the consecutive state of S_1. That is, when the interpreter program interprets at time t_1 $condition_1$ of S_1 to 1 and executes the control action $action_1$, then it is expected that $condition_2$ will be interpreted to 1 at some time $t_2 > t_1$.

Let us assume now that S_2 has VOA. This implies that when $condition_1$ is already interpreted to 1 and $action_1$ is executed, the intended effect of this control action upon the plant described in terms of $condition_2$ will not be achieved. Therefore, $condition_2$ will not be interpreted to 1 at time t_2 but instead the condition of some other controller state, say S_o will be interpreted to 1. The explanation for this situation is that there are a number of assumptions about the plant and about the effects of control actions upon the plant which are not stated explicitly in the control specification and in turn are not represented explicitly in $condition_2$. Thus, when these assumptions are not valid any more $condition_2$ ceases to represent any physically possible plant output, hence it cannot be interpreted to 1.

We would like to stress here that there is a distinction between a state with VOA and an ill-represented state. The condition of an ill-represented controller state is interpreted at some time point to 1 with outdated plant signals and thus, does not represent the actual plant state at this time point. Since, a control action can realize its effect upon the plant only if its condition represents the actual plant state, the control action of the ill-represented state will not realize it's expected effect upon the plant. Thus, the expected condition of the state consecutive to the ill-represented state will not be interpreted to 1. In contrast to this, the control action of the state before the state with VOA does not realize its effect on the plant in terms of the condition of the state with VOA since, this condition does not represent any physically possible plant state. However, both cases have the effect that an expected condition does not interpret to 1, but instead some other condition interprets to 1. The application program is not aware of the fact that a condition is interpreted with outdated plant signals, nor is it aware of the fact that the ontological assumptions implicit in a condition do not hold anymore. Thus, it cannot be aware of the reason for why the expected condition does not interpret to one.

As in the case of ill-represented states, a state with VOA cannot be located from the syntactical description of goal paths consisting of *<condition, control action>* pairs, nor can it be discovered by using extra timing information. The reason is that expectations about the behavior of a the real plant under control, described syntactically by means of the application program are not met in reality although the syntactical description is correct. Therefore, the prevention problem related to a problematic control situation due to violations of ontological assumptions cannot be solved by just using the syntactical description of the application program. However, the syntactical description can be used to find a solution to the identification and recovery control problems related to a problematic control situation due to violation of ontological assumptions.

2.3.3 The Control Problems Considered In The Book

As already explained we are concerned with the behavior of the application program in a problematic control situation. The application program of a modern PC consists of goal paths where some of the control actions in the controller states of these goal paths correspond to continuous feedback loops while a goal path as a whole realizes sequential control. Since the causes for a problematic control situation can be traced to both the continuous and sequential part of a modern PC we will now describe how the four control problems are treated in the existing practice of control applications based on modern PCs.

The continuous feedback control part of a PC

Traditional continuous control applications can be integrated into the PC architecture as follows:

- *Direct implementation.* Direct implementation of a discrete algorithm such as the *direct digital control* (DDC) algorithm [4] using the standard PC language.

- *PC feedback control library.* A modern PC has normally a high level programming language, such as a *functional block diagram language.* The language has built-in a library of function blocks (elements) for feedback control applications such as three-state, PI, PDP, PIP or PID controllers, adaptive controllers, fuzzy controllers, etc. These function elements have well-defined features that make them suitable for a large range of applications and can cope with the usual feedback control problems met in industrial applications. Common features are (*i*) several parameter sets; (*ii*) several modes of operations with: bumpless change-over, manual / automatic control modes, auto-tuning, tracking, anti-reset windup, warm / cold start-up, etc. The modes of operation and different control features are enabled or disabled using logical inputs to the control elements. Therefore, for a sequential program a feedback control element is like any other control action: its mode of operation and the control features are enabled / disabled when conditions are true, according to the same *<condition, control action>* paradigm as for any other sequential PC element.

- *Special-purpose controllers.* Special-purpose controllers from third-party manufactures that have specified properties and which are included as control elements in the control network. Regulators are available for applications such as: electrical motor control, temperature regulators, different linear and circular positioning controllers, emission control for fossil fuel engines, etc. These controllers can use internally a variety of control principles (continuous, digital, fuzzy, neural nets). However, besides that, they have a standard interface to the PC controllers: the modes of operation are enabled / disabled with logical inputs and they are integrated into the overall control scheme using the same *<condition, control action>* control paradigm.

Normally during the design phase of a PC based control application, one solution is favored among those shown above. Then, the solution is tested and adjusted independently of the overall PC control application. If problems appear at some pre-commissioning stage, they can be solved by experts. At the earlier stages before commissioning, the costs related to isolated continuous feedback regulator problems are low compared to the cost for the tests performed during commissioning when these tests are performed on the overall control system. Therefore, during commissioning and delivery of the overall application, the four control problems related to a problematic control situation with causes stemming from the continuous feedback part of the PC are supposed to be already solved. These observations are supported by results reported in [2]. Thus, we assume in this book that the only causes for a problematic control situation are the ones for sequential control.

The sequential control part of a PC

(1) Reachability-related causes.

A number of sequential control related control problems can be normally detected at the syntactical level of the application program. For example the standard IEC 1131 ([7] pp. 183) gives syntactical rules which prevent unsafe transitions in a sequential program, unreachable branches and lock-up. Each PC performs normally a syntactical analysis of the application program and detects such causes. However, there are causes which normally are not represented in the PC interpreter, but which can be described and prevented with formal program analysis, for example using a Petri net representation of the application program. This guarantees that the following causes for a problematic control situation do not appear and are not considered in this book:

- Instability (the state convergence of the plant -- i.e. of the dynamical system -- is guaranteed)

- Incorrect use of resources (the mutual exclusion property is guaranteed)

- Incorrect event ordering (the consistency property is guaranteed)

- Non desirable dynamic behavior (no deadlock, live lock properties are guaranteed)

- Uncoordinated tasks (coordination of tasks property is guaranteed)

In a nutshell, there are available methods that give a solution to the description and the prevention control problems related to a problematic control situation due to the above listed causes. In practice, a standard PC automates few of these methods. Some methods can be used off-line, for instance describing the application program in a Petri net formalism and verifying that the above listed properties hold. A PC has several ways of dealing with the recovery control problem after a sequential control failure. The most important is the jump of the program interpreter, the existence of a hierarchical control level which detects goal path inconsistencies in parallely executed control programs, and the availability of timing operations on parallely executing tasks. The jump function is available with ladder diagrams as well and corresponds to the Grafcet *state forcing*. The recovery principles used in applications are mostly ad-hoc.

(2) Timing causes

The problems of this instance bear high programming, test and installation costs for applications with PCs. Most often, complex control applications with PCs have no formal proofs that the timing constraints are fulfilled. However, several methods are available to enforce time constraints at run time and a large number of tests are performed manually during control system commissioning to ensure that the timing requirements are fulfilled. Normally a PC firmware has several functions that can enforce time constraints such as watchdog timers and synchronization signals among parallel control programs. An application watchdog timer is a down-counter which triggers an alarm when a pre specified event in the controlled process does not occur within a given time interval. Besides application watchdog timers, each control program (task) in a PC has by design a system watchdog timer which ensures that the real-time properties of each control program are fulfilled. There are also available standard synchronization mechanisms for application programs such as those described in [7] page 222, which ensure that tasks act simultaneously.

The recovery control problem related to a problematic control situation due to timing, is a difficult problem and there are no general purpose methods used in industry. Normally a PC uses ad-hoc, application-dependent solutions. In applications the recovery control actions are restricted to hot initializations, forcing states or closing low priority tasks. The recovery control problem for timing failures is a state synchronization problem. Normally there is no difference between the recovery control problem when a problematic control situation is encountered due to reachability and / or due to timing errors.

(3) Ill-represented states causes

Solving the four control problems related to a problematic control situation due to ill-represented states is essential for safety critical applications. For industrial control applications with PCs there are no available general methods for solving the identification and recovery control problems related to a problematic control situation due to ill-represented states. This means that, if a condition though interpreted to 1 does not represent the actual state of the plant under control, then the PC cannot detect automatically this situation and cannot recover from it, when it occurs.

The industrial practice to avoid problems due to ill-represented states is to manually verify that each goal path can be fully interpreted under conditions which may lead to ill-represented states. This procedure has clear drawbacks: a manual test implies that several simplifications are made for the plant and the timing constraints. The real plant may behave differently or the tests may not cover all the possible cases. Most often tests are not performed to simulate the repercussions of the possible existence of a problematic control situation due to ill-represented states.

(4) Violation of ontological assumptions causes

A problematic control situation due to violation of ontological assumptions is essential for safety critical applications. There are no general methods for solving the four control problems related to a problematic control situation due to violation of ontological assumptions. Although violation of ontological assumptions is a frequent cause for a problematic control situation in industrial applications, and even though it can create serious safety problems, there are no methods available which can differentiate this particular cause from the other causes for a problematic control situation and especially from a cause such as ill-represented states.

2.3.4 Summary

Table 2 presents a summary of how complex control applications typically cope with the four control problems related to a problematic control situation due to the different causes from above.

In the context of the above table, the major objective of this book is to propose solutions to the description, identification and partly, the recovery control problems when a problematic control situation is encountered due to violations of ontological assumptions (shown with shadowed background in Table 2). The focus is on the identification and recovery control problems, namely how can a controller be designed such that it detects a violation of ontological assumptions in real-time and is able to recover from such a violation? A controller with this property is called an ontological controller and its architecture and design principles are presented in Chapter 7.

Table 2

Type of control	The Description Control problem	The Prevention Control Problem	The Identification Control Problem	The Recovery Control Problem
(1)Continuous control .	Well-known solutions	Modeling, specific control laws	Manual or automatic supervision	Auto-tuning, manual adjustment,adaptations
Sequential control: (2) Reachability-related causes	Solutions not as well-known as in (1)	PC syntax check, off-line sequential modeling	Manual; watchdog timer supervision	Ad-hoc. Controlled initialization or synchronization operation
(3) Timing-related causes	Solutions less known than in (2)	Manual tests	Manual; watchdog timer supervision	Ad-hoc. Controlled initialization or synchronization operation
(4) Ill-represented states and VOA	Almost unknown	May be discovered during manual tests	Unknown	Unknown

The ontological controller supervises an object controller, that is a particular application program and thus, it interacts with the components of the application program that are essential for its real-time behavior such as the goal seeking and the synchronization operations which in turn use the state set of the object controller to perform their respective functions. The goal seeking operation and the synchronization operation are relevant since they are instrumental for solving the recovery control problem in the case of a problematic control situation due to causes different from a violation of ontological assumptions. The goal seeking and the synchronization operations are formally introduced in Chapter 3 and the interaction between the ontological controller, the goal seeking and synchronization operations is described in Chapter 6.

A major result shown in Chapter 3 is that for an unrestricted controller state set the goal seeking and the synchronization operations as defined in this chapter cannot distinguish the different causes for a problematic control situation from each other.

A major result shown in Chapters 4 and 5 is that if the controller state set is restricted in an appropriate manner (shown in Chapter 4), then violations of ontological assumptions can be distinguished from the following types of PCSs: unexpected external actions, control time de-synchronizations and ill-represented states (shown in Chapter 5).

A major result in Chapters 6 is a prove that a certain goal seeking operation together with a non-well determined state set can exhibit the same control properties as a well-determined state set. Consequently violations of ontological assumptions can be detected even for non-well-determined state sets by properly constraining the goal seeking operation.

A major result in Chapter 7 is a control architecture that uses the results from Chapter 6 to determine a special well-determined state set starting from a non-well-determined state set and a constraint GSO. This state set, together with the results from Chapter 5, show that a controller using this architecture can detect violations of ontological assumptions. With this, the declared goal of the ontological control is accomplished.

Yet, there are many unanswered questions. Some of these are summarized under the last section about further research topics.

REFERENCES

[1]   *** Asea Brown Boveri Corporate Research. *Computer Science 1993 Acivities. Overview.* November 1993.

[2]   Benveniste, A., Åström, K. J., Caines, P.E., Cohen, G., Ljung, L., Varaiya, P. Facing the challenge of computer science in the industrial application of control: a joint IEEE CSS-IFAC project. In IEEE Transactions on Automatic Control vol. 38, No.7 July 1993.

[3]   Bennett, M.E. Real-time continuous AI systems. In IEE Proc. vol 134, Pt. D., No. 4 July 1987.

[4]   Faurre, P., Depeyrot M., - *Elements of System Theory*, North-Holland Publishing Company, Amsterdam, New York, Oxford, 1977.

[5]   *** International Electrotechnical Commission. International Standard. Programmable Controllerrs Part 1: General Information. IEC 1131-1.

[6]   *** International Electrotechnical Commission. International Standard. Programmable Controllerrs Part 2: Equipment Requirements and Tests. IEC 1131-2.

[7]   *** International Electrotechnical Commission. International Standard. Programmable Controllerrs Part 3: Programming Languages. IEC 1131-3.

[8]   *** *Knowledge-Based Real-Time Control Systems. IT4 Feasibility Study,* ABB, SattControl AB, TeleLOGIC AB. Department of Automatic Control, Lund Institute of Technology, Studentlitteratur 1988, Sweden.

[9]   Inger Klein, *Automatic Synthesis of Sequential Control Schemes.* PhD thesis. Dept. of Electrical engineering, Linköping, Sweden 1993.

[10]   Lin, Jing-Yue, Ionescu, Dan. A generalized temporal logic approach for control problems of a class of nondeterministic discrete event systems. Proc. of the 29-th IEEE Conf. Decision and Control, Honolulu, Hawai, Dec. 5-7, 1990, pp 3440-3445.

[11]   Feng Lin, Analysis and Synthesis of Discrete Event Systems Using Temporal Logic. In Proceedings of the 1991 IEEE Intnl Symposium on Intelligent Control, 13-15 August 1991, Arington Virginia, USA.

[12]   Fangzhen Lin, Shoham, Y., Provably Correct Theories of Action (preliminary report), in: National (U.S.) Conference on Artificial Intelligence, pages 349-352, 1991.

[13]   Olsson, G., Piani, G. *Computer Systems for Automation and Control.* Prentice Hall, New York, 1992.

[14]   Reising, W. *Petri Nets, An Introduction.* EATCS Monographs on Theoretical Computer Science vol. 4, Editors W. Brauer, G. Rozenberg, A. Salomaa. Springer-Verlag Berlin, Heidelberg, New York, Tokyo, 1982.

[15]   Ravn, A.P., Rischel, H., Hansen K.M. Specifying and Verifying Requirements of Real-Time Systems, IEEE Trans on Software Eng. January 1993

[16]   Ramadge, P.J., Wonham, W.M., Supervisory control of a class of discrete-event processes, SIAM Journal, Control Optimization, vol 25, no.1, pp. 206-230, Jan. 1987.

[17]   Ramadge, P.J.G., Wonham, W. M. The Control of Discrete Event Systems. In Proc. of the IEEE, vol 77, no.1 Jan 1989, pp. 81- 97.

[18]   Thistle, J.G., Wonham, W.M., Control problems in a temporal logic framework. In Int. J. Control vol. 44 no. 4 pages 943-476.

[19]   Wei Li, Towards a Theory of Requirement Capture. Dept. Of Computer Science, Beijing University of Aeronautics and Astronautics, P.R. China.

[20]   Willner, Y., Heymann, M. On language convergence in discrete-event systems. Proc. 17-th Convention of Electrical and Electronics Engineers in Israel, 1991.

[21]   Wonham, W. M., A control theory for discrete event systems. In Advanced Computing Concepts and Techniques in Control Engineering, M.J. Denham, A.J. Laub, editors, pp. 129-169. Springer-Verlag, 1988.

[22]   Zhong, H, Wonham, W.M. Hierarchical Coordination. IEEE 1990 pages 8-14.

[23]   Åström, K.J., Hägglund, T., Automatic Tuning of PID Controllers. Instrument Society of America, Research Triangle Park, NC27709.

3

FORMAL DESCRIPTION

In Chapter 3 we introduced informally the basic concepts involved in control with a PC such as controller states, goal states, and goal paths. Furthermore, we described the two operations of synchronization and goal seeking which a PC uses in the case of a problematic control situation as well as a number of causes for such a problematic control situation. Let us remind here that a problematic control situation was identified relative to a control de-synchronization, i.e., a jump from a currently interpreted goal path to a new goal path.

In this chapter we describe formally the concepts of controller state, goal state, and goal path, by introducing formal constructs such as plant formulas, control actions, and state transitions. Then we consider a problematic control situation in terms of a de-synchronization from a currently interpreted goal path. A de-synchronization is defined as a transition from the currently interpreted goal path to the same or another goal path whose transition is not among the state transitions due solely to control actions. We also describe formally the causes for a de-synchronization in terms of expected and unexpected external actions, ill-represented states, and violation of ontological assumptions. Furthermore we provide a formal description of the goal seeking and synchronization operations and show their formal properties in the case of de-synchronization due to particular causes.

The major objective of the above described formalization of control with a PC is twofold:

- To determine if the PC exhibits some characteristic behavior in the case of a de-synchronization due to violation of ontological assumptions, and

- To determine whether the behavior of the PC in the case of de-synchronization due to violation of ontological assumptions can be distinguished from its behavior in the case of de-synchronization due to other causes such as expected, unexpected external actions, and ill-represented states.

To this end we generalize the behavior of the GSO in terms of different types of transition sequences where each type of a transition sequence is due to combinations of different causes for de-synchronization. We conclude by describing the ways for distinguishing between the different types of GSO behavior and the causes for it.

3.1 Basic Notions And Definitions

This section describes in formal terms the basic concepts involved in control with a PC such as: plant formulas, control actions, controller states, state transitions, goal states and goal paths. It also introduces the formal description of two particular causes for a problematic control situation, namely expected and unexpected external actions. We conclude with a description of the types of control knowledge used in applications for the construction of the above mentioned formal concepts.

The control scheme with PCs, shown in **Figure 9**, consists of a *plant*, i.e., the well defined part of the process under control and a PC. The latter acts upon the plant by executing a sequence of control actions, where each control action produces a particular plant output. External actions upon the plant are executed independently of the control actions of the PC and they also result in particular plant outputs.

3.1.1 Plant Outputs

It is assumed here that the output of the plant under control can be described by a finite number of distinct states. These states are normally derived from the plant model and additional unmodelled knowledge about the plant (e.g., heuristics, specification, experiments, etc.).

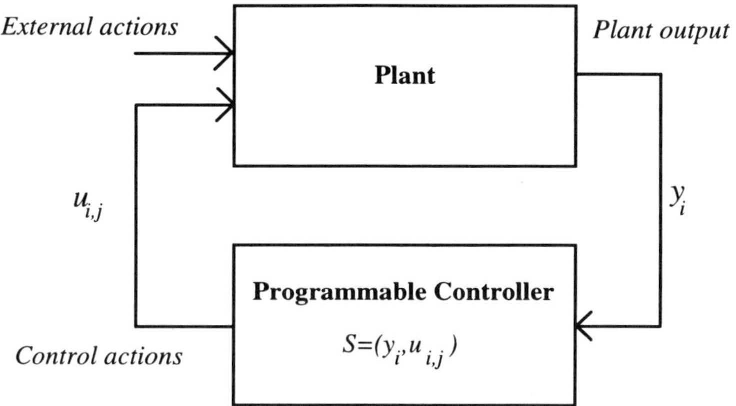

Figure 9 Control Scheme with Programmable Controller.

Each plant output is expressed by the so-called *plant-formula*. A plant-formula is *interpreted* at a *sampling instant, t_k*, where $t_k = k \cdot h$, k being an integer, and h being certain sampling period. Thus, we consider the case of periodic sampling (i.e., the sampling instants are equally spaced in time). For the purpose of our presentation we will denote a sequence of sampling instants $(t_k, t_{k+1}, ...,t_{k+n})$ by $(t, t+1, ...,t+n)$ where the transformation from the first to the second sequence is done by substituting t_k with t, t_{k+1} with $t+1$, etc.

The interpretation of a plant-formula results in it being evaluated as *true (1)* or *false (0)*. When the plant output expressed by the plant-formula is observed (holds) at t, the plant formula is interpreted to *1*, otherwise it is interpreted to *0*.

In the following we will describe the construction of a plant-formula and how such a formula is interpreted.

3.1.2 Plant Signals

A *plant signal*, denoted as *x*, represents a physically measurable internal plant-variable, or is an estimate (obtained via an observer) of an internal plant variable. Normally, the physically measurable internal variables constitute the output of the plant and are a subset of all internal variables.

When *x* corresponds to a physically measurable plant output, i.e., it is provided by a sensor, it is very often the case that *x* is a function of some other signals, called *primitive signals*. For example, a *Pt100* temperature sensor provides a temperature measurement (a plant signal) which is obtained as a non-linear function of current (a primitive signal). In other cases, the function representing x may also include time as one of its arguments ([1]).

The primitive signals are usually constrained due to sensor limitations and actuator properties. A constraint on a primitive signal defines its normal range of variation. Consequently, a normal range of variation for a plant signal is derived by taking into account the normal variation ranges of primitive signals and the function describing the plant signal in terms of these primitive signals.

Different variation ranges for the same plant signal may be defined for normal operation in different control modes, or for operation during some critical conditions (e.g., alarms), sensor failure, etc.

The plant signal x is measured or estimated at sampling instant t and its physical value, denoted as $x(t)$, is called the *interpreted value* of x at sampling instant t.

We consider the finite set X of n plant signals $\{x_1,...,x_n\}$. Thus, $X(t) = \{x_1(t),...,x_n(t)\}$ is the set representing the physical values of all interpreted plant signals at sampling instant t.

3.1.3 Constrained Plant Signals

A plant signal, x, may be further constrained within its range of variation for the purpose of specifying the conditions under which the execution of a control action can be expected to result in a particular plant output as well as for specifying the expected plant output itself.

A *constrained plant signal* or simply a *plant constraint* on x_i, denoted as v_i, is a binary relation ($\leq$, $\geq$, or $=$) between x_i and certain physical value, or a function of physical values. There is normally a set of plant constraints associated with each x_i. This set is denoted as $V_i = \{v_{i,1}, v_{i,2},..., v_{i,m}\}$.

Each plant constraint $v_{i,j}$ is interpreted at sampling instant t that is, it is verified whether the binary relation it represents holds or not at t for the interpreted value $x_i(t)$ of x_i at t. Thus, interpreting the relation $v_{i,j}$ means that at sampling instant t it is assigned a value $v_{i,j}(t)$ such that, $v_{i,j}(t) = 1$ if the relation holds for $x_i(t)$, and $v_{i,j}(t) = 0$ if the relation does not hold for $x_i(t)$. In this context, each $v_{i,j}$ is a boolean variable expressing a constraint on a plant signal x_i. The set $V_i(t) = \{v_{i,1}(t), v_{i,2}(t), ...,$

$v_{i,m}(t)\}$ represents the interpreted values for all primitive constraints on the plant signal x_i at sampling instant t.

Furthermore, $V = V_1 \cup V_2 \cup ... \cup V_{im}$, is the set of all plant constraints on all plant signals and consequently, $V(t) = V_1(t) \cup V_2(t) \cup ... \cup V_{im}(t)$ is the set of all interpreted values for all plant constraints at sampling instant t.

3.1.4 Plant Formulas

A *plant formula* represents a plant output, y, and is defined as a boolean function of plant constraints, $v_{i,j}$, on directly measurable and/or estimated plant signals x_i.

A plant formula is interpreted at sampling instant t by computing a value for the boolean function given certain $V(t)$. Thus, an *interpreted plant formula*, y, takes the value $y(t)$ which is equal to 1 when the plant output represented by y is physically present, and $y(t)$ is equal to 0 when the plant output is not physically present.

The plant formula y_i is *subsumed* by the plant formula y_j if and only if whenever $y_j(t) = 1$ then $y_i(t) = 1$, but not vice versa. We write $y_i \subset y_j$ to denote that y_i is subsumed by y_j.

We consider the finite set Y of all plant formulas where, $Y = \{y_1, y_2, ..., y_M\}$. Thus, $Y(t) = \{y_1(t), y_2(t), ..., y_M(t)\}$ is the set of interpreted plant formulas.

For the purpose of this presentation, we will only use the symbols y_i, denoting plant formulas and the symbols $y_i(t)$, denoting interpreted plant formulas, without the explicit form of the boolean functions behind these symbols.

3.1.5 Plant Formulas Of A Well Designed Object PC

An *object PC* is a PC which performs a control task alone, or when it is included in a control hierarchy, it is at the control level closest to the plant. For example a particular PC in the networked architecture shown in **Figure 5** is an object PC. We shall consider as object PCs only the so called *well-designed object PCs*. A well-designed object PC is characterized by certain properties of its plant formulas and control actions. The plant formulas of a well-designed object PC have the following properties:

- At each sampling instant t exactly one plant formula y_i is interpreted, that is, $y_i(t)=1$ or $y_i(t)=0$

- Plant formulas are not subsumed: there do not exist two plant formulas y_i and y_j such that $y_i \subset y_j$

- The set Y of plant formulas is defined a priori and is not modified during control by adding new, or deleting existing plant formulas

In the rest of the book a well-designed object PC will be referred to as an object PC .

3.1.6 Control Actions And External Actions

The purpose of an object PC is, using available interpreted plant formulas, y, to determine *control actions* (control signals) u, so that its control goals are eventually achieved. Normally, a control action is performed on the basis of a feedback from the plant output, but other actions (*external actions*), outside of the control loop, may also influence the plant, e.g., manual operator commands, actions performed on the plant by objects and / or agents in its environment.

Control actions

When a PC executes a control action, it changes the current plant output, i.e., the current plant formula is transformed to itself or to another plant formula. For a particular control action, u, to produce certain desired change (control *action post-condition*) in the plant output at $t+1$, certain plant formula (*control action precondition*) has to be interpreted to true at t. For example, let the action u be such that it can transform y_3 into y_4. Now, if u's precondition y_3 is interpreted to true at a sampling instant t and u is executed at t, it is expected that y_4 will be interpreted to true at the next sampling instant $t+1$.

However, for the physical execution of a control action it is not necessary that the plant formula describing its precondition has to be interpreted to true at t. When the control action precondition does not hold at t but the control action is executed anyway, its post-condition will not be interpreted to true at $t+1$. Also, it may not always be the case that the plant formula expressing the control action post-condition will be interpreted to true at $t+1$, despite of the fact that its precondition was interpreted to true at t.

One and the same physical control action u may have more than one precondition and for each precondition it may have only one expected post-condition. To reflect this property of control actions in our formalism we will describe a single physical control action u by a set of distinct symbols $u_{i,j}$ where:

> (i) the index i denotes the i-th plant formula used as precondition of the control action u, and

> (ii) the index j denotes the j-th plant formula used as the expected post-condition. of u.

Thus, $u_{i,k}$ and $u_{m,j}$ become two distinct notations for the same physical control action u where: (i) $u_{i,k}$ describes u in the context of a precondition y_i and an expected post-condition y_k, and (ii) $u_{m,j}$ describes the same u in the context of a different precondition y_m and a different expected post-condition y_j. It is easily seen that for example, $u_{i,k}$ and $u_{i,j}$ must correspond to two different physical control actions u and u' since they have the same precondition, namely y_i. This is so because one and the same physical control action when executed under the same preconditon cannot result in two different plant outputs, i.e., y_k and y_j. In a similar manner, $u_{j,k}$ and $u_{i,k}$ must correspond to two different physical control actions since one and the same physical control action when executed under different preconditions cannot result in the same plant output, i.e., y_k.

The set of all the physically possible control actions of a PC is denoted $U=\{u_{i,j}\}$ $i=1,...,M; j=1,...,M$.

We say that a control action $u_{i,j}$ is interpreted to true at t (or executed at t) if the physical control action u corresponding to $u_{i,j}$ is actually executed at t. Also, $u_{i,j}$ is interpreted to false at t if the physical control action u corresponding to $u_{i,j}$ is not executed at t. An interpreted control action is denoted as $u_{i,j}(t)$ and as a result of its interpretation (execution), we may have that $u_{i,j}(t) = 1$, in case $u_{i,j}$ is interpreted to true at t, or $u_{i,j}(t) = 0$, in case $u_{i,j}$ is interpreted to false at t.

The set of control actions of an object PC, has the following characteristics:

- For each plant formula y_i there exists at least one control action $u_{i,j}$ such that at any t, when $y_i(t)=1$ then $u_{i,j}(t)=1$.

- If $y_i(t)=0$ and $u_{i,j}(t)=1$ then $y_j(t)=0$ (i.e., a control action does not have the expected effect upon the plant if its precondition is interpreted to false at t and the control action is executed at t).

- If $y_i(t)=1$ and $u_{i,j}(t)=1$ then it is **expected** that $y_j(t)=1$ (i.e. a control action is expected to have a pre-specified effect upon the plant effect only if its precondition is interpreted to true at t and the control action is executed at t).

- At each t exactly one control action $u_{i,j}$ is executed (interpreted to true), that is: $u_{i,j}(t)=1$.

- The set of control actions U is determined a priori and is not modified during control by adding new control actions or deleting existing control actions.

External actions

As already mentioned in the beginning of this section, other actions (external actions), outside of the control loop, may also influence the plant output, e.g., manual operator actions, actions performed on the plant by objects and/or agents in its environment, changes in plant output because of the occurrence of some expected and unexpected events in the environment, etc. Thus, these external actions may also result in a transformation of plant formulas to other plant formulas.

There are two types of external actions:

- expected external actions

- unexpected external actions

In what follows we will briefly describe the above two types of external actions.

Expected external actions

An *expected external action*, u^{ext}, can be described (in terms of pre- and post-conditions) if for example, it belongs to the set of allowable operator actions, or represents an event that can be expected to occur during control, and affects the plant output in a known way. However, the exact time when such an action may take place is not known in advance. Thus, the precondition of u^{ext} may be any plant formula which happens to be interpreted to true at a certain time t. As a consequence of this, the post-condition of u^{ext} has to be determined for each plant formula $y_i \in Y$. In this context, u^{ext} is denoted as $u_{i,j}^{ext}$, $i, j = 1,...,M$.

Another feature of u^{ext} is that it can be interpreted to true, **only** at a later sampling instant than the sampling instant at which it actually occurred. Suppose that at t we have that the plant output is represented by the plant formula y_i, that is $y_i(t) =$

1 and let there be a control action, $u_{i,j}$ which given that $y_i(t) = 1$ is expected to result in a new plant formula y_j, i.e., it is expected that $y_j(t+1) = 1$. Suppose also that at t the controller executes $u_{i,j}$, but at $t+1$ the plant formula interpreted to true is y_k instead of y_j, i.e., we have that $y_k(t+1) = 1$ and $y_j(t+1) = 0$. Furthermore, let there be an expected external action $u_{i,k}^{ext}$, that is an expected external action which when it occurs, given that $y_i(t) = 1$, results in $y_k(t+1) = 1$.

The situation described above indicates that an expected external action has occurred at t which has transformed the plant output to y_k. Furthermore, this action has overridden the expected effect y_j of the control action $u_{i,j}$. However, we are able to identify the occurrence of $u_{i,k}^{ext}$ at $t+1$ rather than at t when the action actually occurred.

In conclusion, an expected external action for an object PC has the following properties:

- It is defined a priori and denoted as $u_{i,j}^{ext}$, $i=1,...,M$ and $j=1,...,M$ and its occurrence is **expected** to transform a plant formula y_i into a plant formula y_j.

- An expected external action can occur at any time instance t, but its occurrence can be identified only at the next sampling instant $t+1$.

Unexpected external actions

An *unexpected external action* cannot be defined in terms of its pre- and post-conditions. However, its occurrence can be detected in the following way.

Suppose that at t we have the plant formula y_i, that is $y_i(t)=1$ and let there be a control action, $u_{i,j}$ which given that $y_i(t)=1$ is expected to transform y_i into a new plant formula y_j, i.e., it is expected that $y_j(t+1)=1$. Suppose also that at t the controller executes $u_{i,j}$, but at $t+1$ we obtain y_k instead of y_j, i.e., we have that $y_k(t+1)=1$ and $y_j(t+1)=0$. Furthermore, let there be no expected external action $u_{i,k}^{ext}$, i.e., when $u_{i,k}^{ext}$ occurs, given that $y_i(t) = 1$, it is expected to result in $y_k(t+1) = 1$.

The above described situation is an indication that an unexpected action has occurred which has overridden the expected effect y_j of the control action $u_{i,j}$. However, the above described situation may be also the result of violation of ontological assumptions. In this context, one major goal of ontological control is to provide the means for distinguishing between the occurrence of unexpected external actions and a violation of ontological assumptions.

In conclusion, the unexpected external actions of an object PC have the following properties:

- An unexpected external action does not have a corresponding notation.

- An unexpected external action can occur at any time instance t;

- The occurrence of an unexpected external action results in a plant formula which belongs to the a priori defined set Y of plant formulas.

3.2 Controller States And State Transitions

As already described in Chapter 2, the application program of an object PC can be appropriately described in terms of controller states (*<condition, control action>* pairs), and goal paths (sequences of controller states leading to a goal state). In this section we provide a formal description of controller states in terms of the plant formulas and control actions from the previous section, and, we define formally goal paths in terms of state transitions between controller states.

We also state the properties of the controller states and the goal paths of an object PC.

3.2.1 A controller State

A *controller state*, or simply a *state* denoted as S, is an ordered pair (y, u) where $y \in Y$ and $u \in U$. Thus, a state specifies the control action to be executed given the presence of a particular plant output.

A well-defined controller state

A *well-defined controller state* or simply a *well-defined state*, S, is represented by the ordered pair $(y_i, u_{i,j})$ or by the ordered pair $(y_i, u_{i,j}^{ext})$. In other words, when the control action $u_{i,j}$ is executed at t, given that its precondition y_i is interpreted to true at t, it is expected that the control action post-condition y_j will interpret to true at $t+1$. In the same fashion, when the expected external action $u_{i,j}^{ext}$ occurs at t, given that its precondition y_i is interpreted to true at t, it is expected that its post-condition y_j will interpret to true at $t+1$.

The set of well-defined controller states defines the *state set* of the object PC. In what follows we will consider only a state set consisting of well-defined states, that is, states of the form $(y_i, u_{j,k})$ or $(y_i, u_{j,k}^{ext})$ do not belong to the state set of an object PC. The state set of an object PC is *ontologically complete* if and only if for each y_i there is at least one control action $u_{i,k}$ associated with it.

A *controller partial state*, or simply a *partial state* is a state of the type $(y_i, -)$, where "-" means that the control action part of the state is not yet identified with any particular control action but, if needed it can be completed with any one of the control actions $u_{i,j}$, $i, j = 1,...,M$. Thus, the completion of a partial state always results in a well-defined state.

An interpreted well-defined controller state

An interpreted well-defined controller state, or simply an interpreted state, denoted as $S(t)$ is given by the ordered pair $(y_i(t), u_{i,j}(t))$ where, $y_i(t), u_{i,j}(t) \in \{0, 1\}$. That is, the control action and its precondition constituting a well-defined controller state are interpreted to either true or false at t. Thus, we can have a number of cases: (1) the precondition was observed to hold at t, but the control action was not executed at t; (2) the precondition was observed not to hold at t, and the control action was not executed at t; (3) the precondition was observed not to hold at t, and the control action was executed at t; (4) the precondition was observed to hold at t, and the control action was executed at t.

An *interpreted well-defined partial state*, or simply an *interpreted partial state*, denoted as $S(t)$ is the state $(y_i(t), -)$, where $y_i(t) \in \{0, 1\}$.

A material well-defined controller state

A material controller state, or simply a material state, is an interpreted state, $S(t)$ such that, $S(t) = (y_i(t), u_{i,j}(t))$ and $y_i(t) = 1$ and $u_{i,j}(t) = 1$. That is, the control action and its precondition are both interpreted to true. In other words, the control action is executed at t and its precondition was observed to hold at t. When a state is interpreted so that it becomes a material state we say that the state in question is materialized.

A material partial state is an interpreted partial state $(y_i(t), -)$ where, $y_i(t) = 1$.

A goal controller state

A *goal controller state*, or simply a *goal state*, denoted as G, is a state $(y_i, u_{i,i})$. The control goal of an object PC is to materialize a goal state.

A control action $u_{i,i}$, according to the notation conventions in Section 3.1.6, transforms a plant formula y_i into itself. Therefore, an action symbol with two identical indices, $u_{i,i}$, corresponds to the empty (null) physical control action, which does not have any effect upon the plant.

The set of goal states has the following characteristic features:

- The set of goal states is a subset of the state set S.

- A state is a goal state iff it has a null action, $(y_i, u_{i,i})$ (i.e., each goal state has a null control action and reciprocally, each state in the state set which has a null action is a goal state).

- The goal states are totally ordered in terms of their priority, i.e., the control specification for an object PC defines the order in which the goal states are to be materialized. The priority relation between goal states is denoted '>'. The notation $G_a > G_b$ means that the goal state G_a is specified to materialize prior to G_b.

- The set of goal states is determined a priori and is not modified during control by adding new goal states or deleting existing goal states.

- A goal state is achieved if and only if the plant formula component of the goal state is interpreted as true (materialized) at certain t.

3.2.2 Controller State Transitions

Controller state transitions, or simply state transitions, take place between a particular type of well-defined states called consecutive states.

Consecutive states

The S' is called the *consecutive state* of another state S if and only if $S = (y_i, u_{i,j})$ and $S' = (y_j, u_{j,k})$. That is, the $u_{i,j}$'s post-condition y_j is the same as the control action precondition from S'. Normally each state has a set of consecutive states, e.g., the set of consecutive states of $S = (y_i, u_{i,j})$ is the set of states $\{(y_j, u_{j,k})\}, k \in \{1, ..., M\}$.

The material state, $S'(t)$ is called the *consecutive material state* of another material state $S(t)$ if and only if $S(t) = (y_i(t), u_{i,j}(t))$ and $S'(t+1) = (y_j(t+1), u_{j,k}(t+1))$. That is, the control action from the state S when executed (interpreted to *1*) results in the control action precondition from S' being interpreted to *1* and the consequent execution of the control action from S'. Any material state at t has only one consecutive material state since only one plant formula and only one control action can be both materialized at $t+1$.

A controller state transition

A *controller state transition* or simply *state transition*, is defined as the transition from one well-defined state to one of its consecutive well-defined states. Thus, a state transition can **only** be due to a control action.

A *controller transition* or simply a *transition* is the transition from a state to another state which is **not** the consecutive state of the first one. Such transitions are due to the occurrence of expected external action, or unexpected external action etc.

A state transition is denoted as $S \rightarrow S'$ where S' is a particular consecutive state of S. A transition is denoted as $S - -> S'$.

A *sequence of controller state transitions* or simply *state transition sequence* can be defined by starting from a particular well-defined state followed by one of its immediate consecutive state, followed further by one of the consecutive states of this immediate consecutive state, etc. In the context of a given state transition sequence we say that a state S' is *reachable* from another state S if and only if there exists at least one state transition sequence for which the first state in the state transition sequence is S and the last state in the state transition sequence is S'.

A sequence which consists of both state transitions and transitions is called a *transition sequence.* In the context of a given transition sequence we say that a state S' is *reachable* from another state S if and only if there exists at least one transition sequence for which the first state in the transition sequence is S and the last state in the transition sequence is S'.

A state transition sequence is called *optimal* if it maximizes/minimizes certain performance criteria, e.g., steady state variance (minimum variance control), state transition time (time optimal control), or an application oriented performance index related to cost, product quality, resource expenditure (e.g., fuel, energy, and raw materials consumption). A transition sequence is **never** optimal since it has uncontrollable expected external actions (only the component state transition sequences can be optimal). In this book, the particular optimality criteria for a state transition sequence is not of interest.

A state transition $S \rightarrow S'$ or a transition $S \dashrightarrow S'$ is interpreted to true (or materialized) if and only if S is material state at t and S' is a material state at $t+1$. A state transition between material states is denoted as $S(t) \rightarrow S'(t+1)$ and is called a *material state transition*. A transition between material states is denoted as $S(t) \dashrightarrow S'(t+1)$ and is called a *material transition.*

Goal paths

A *goal path* is a state transition sequence from some initial state to a goal state. That is, for each state $(y_i, u_{i,j})$ in a goal path we have that $u_{i,j} \in U$. A state belonging to a goal path is called an *image state*. The set of all states on all goal paths is called the *set of image states*. The goal paths are assumed ordered according to an application-specific criteria. The goal path with the highest priority is called *optimal.* There is exactly **one** *optimal goal* path leading to a given goal state.

Two goal paths are different if and only if there is at least one image state which belongs to one of them, but does not belong to the other one. A goal path corresponding to a transition sequence is called a *non-optimal goal path.* There may be more than one non-optimal goal paths leading to a goal state.

Optimal and non optimal goal paths are decided during the design of the object PC, but they are not represented explicitly. Instead, they are implicit in the state set of the object PC.

The image states and goal states of an object PC have the following characteristic properties:

- Repetitions of one and the same image state in the same goal path are not allowed.

- Cycles in and between goal paths are allowed. One and the same plant formula may be in the plant formula part of image states on the same optimal goal path, or on different optimal goal paths. Thus, one and the same goal path can have the image state $(y_i, u_{i,j})$ followed further on in the same goal path by the state $(y_k, u_{k,i})$, which represent a cycle since the first state is reachable from the second one and vice versa

- One and the same image state may belong to more than one optimal goal paths. For example, $(y_i, u_{i,j})$ may be a state that belongs to the optimal goal path leading to the goal state Sg as well as to the optimal goal path leading to the goal state Sg'.

- The last state of a any goal path (optimal or non optimal) is a goal state. A goal state does not have any consecutive image state.

- Any image state S, different from a goal state, has at least one consecutive state. Exactly one of these consecutive states is on the same optimal goal path as S and the rest may be on other goal paths. Thus, the set of consecutive states of an image state is **never** empty.

- The goal paths are ordered with the priority relation '>' as follows:

  - One optimal goal path has a higher priority than another optimal goal path if and only if the first one leads to a goal state G_a and the second one to a goal state G_b, such that $G_a > G_b$.

  - There is no priority order on non optimal goal paths.

- Since the optimal goal paths are decided in advance and are not modified during control by adding new or deleting existing image states, the set of image states is constant. Thus, the set of consecutive states of an image state is also constant.

For example, consider the image state $(y_i, u_{i,j})$ on the optimal goal path to G. Its set of consecutive image states is then the set $\{(y_j, u_{j,k})\}$, $k \in \{1, \dots, M\}$. This set contains at least the image state $(y_j, u_{j,r})$ such that the state transition $(y_i, u_{i,j}) \rightarrow (y_j, u_{j,r})$ is on the optimal goal path to G. However, if the plant formula y_i is in the plant formula part of other image states on the same optimal goal path to G and/or other optimal goal paths, then $\{(y_j, u_{j,k})\}$, $k \in \{1, \dots, M\}$ will further contain more image states from the same optimal goal path to G and/or image states from other optimal goal paths.

When two optimal goal paths (each leading to a different goal state) have image states which have the same plant formula in their plant formula parts, this implies that there are more than one goal paths leading to the either one of two goal states. Let y_j be common for the image states $(y_j, u_{j,r})$ and $(y_j, u_{j,l})$ on the optimal goal paths leading to G and G' respectively (**Figure 10**).

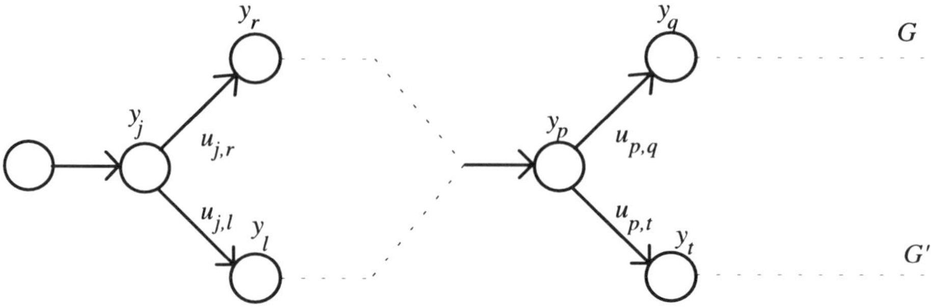

Figure 10 Non-optimal Goal Path.

Furthermore, assume that the object PC has been interpreting the optimal goal path to G and at t, it interprets y_j as true, thus materializing the partial image state $(y_j, -)$. If this partial state is completed with $u_{j,l}$ then the image state, $(y_j, u_{j,l})$, on the optimal goal path to G' will be materialized and the object PC will start interpreting the optimal goal path to G'. Suppose now that these two goal paths have two other image states like $(y_p, u_{p,q})$ on G' and $(y_p, u_{p,t})$ on G, and let at some later $t+n$ we have that $y_p(t+n) = 1$, that is, the partial state $(y_p, -)$ has been materialized. Here, we can complete $(y_p, -)$ with a control action, say $u_{p,q}$ such that the completed state $(y_p, u_{p,q})$ belongs to the optimal goal path to G rather than to the optimal goal path to G'. Thus, there is another goal path leading to G which temporarily deviates from the optimal goal path to this same goal state, e.g. it starts following an optimal goal path to another goal state, G', and finally, returns to the optimal goal path leading to G.

The goal paths leading to the same goal state and different from the optimal goal path to this same goal state are the non optimal goal paths with respect to this goal state. Thus, the goal path to G going through $(y_j, u_{j,l})$ and $(y_p, u_{p,q})$ is non-optimal. If there is both an optimal and a non-optimal goal path leading to the same goal state G the object PC can achieve G, for example, by first following the optimal goal path to G, then following the non-optimal goal path to this same goal state (i.e., a part of an optimal goal path to a different goal state), and finally returning to the optimal goal path to the original goal state G. This can be realized if at least two plant formulas are common for both of these two paths.

To illustrate the above variety of cases, consider the following optimal goal paths:

- $(y_1, u_{1,2}) \rightarrow (y_2, u_{2,5}) \rightarrow (y_5, u_{5,8}) \rightarrow (y_8, u_{8,11}) \rightarrow (y_{11}, u_{11,5}) \rightarrow (y_5, u_{5,14}) \rightarrow (y_{14}, u_{14,30}) \rightarrow (y_{30}, u_{30,30})$

- $(y_3, u_{3,7}) \rightarrow (y_7, u_{7,8}) \rightarrow (y_8, u_{8,11}) \rightarrow (y_{11}, u_{11,5}) \rightarrow (y_5, u_{5,9}) \rightarrow (y_9, u_{9,14}) \rightarrow (y_{14}, u_{14,31}) \rightarrow (y_{31}, u_{31,32}) \rightarrow (y_{32}, u_{32,32})$

- $(y_6, u_{6,8}) \rightarrow (y_8, u_{8,11}) \rightarrow (y_{11}, u_{11,10}) \rightarrow (y_{16}, u_{16,16})$

One can observe here that:

- The plant formula y_5 is part of two image states on the same optimal goal path to the goal state $(y_{30}, u_{30,30})$;

- y_5 is part of image states on the optimal goal path to the goal state $(y_{30}, u_{30,30})$ and it is also part of an image state on the optimal goal path to another goal state $(y_{32}, u_{32,32})$;

- The image state $(y_8, u_{8,11})$ is common for the three different optimal goal paths leading to the goal states $(y_{30}, u_{30,30})$, $(y_{32}, u_{32,32})$, and $(y_{16}, u_{16,16})$.

Furthermore we have that the goal path,

- $(y_3, u_{3,7}) \rightarrow (y_7, u_{7,8}) \rightarrow (y_8, u_{8,11}) \rightarrow (y_{11}, u_{11,5}) \rightarrow (y_5, u_{5,9}) \rightarrow (y_9, u_{9,14}) \rightarrow (y_{14}, u_{14,30}) \rightarrow (y_{30}, u_{30,30})$

is a non-optimal goal path to the goal state $(y_{30}, u_{30,30})$.

Inner state transitions

In practice, the sampling time of the object PC, i.e., the time interval during which it interprets states is small enough so that the object PC can sense and react to any relevant change in the plant output, (i.e. this is the property known as *real-time control*). This means that the sampling time of the object PC is smaller than the time interval during which the plant can change from one output to another. Therefore several interpretations of one and the same plant formula accompanied with several executions of its corresponding control action, may be performed before the plant changes its output. In other words, some states in the goal path are materialized more than once in successive sampling instants.

The situation when a state is materialized a **number** of times and during this, no transition to a consecutive state takes place, is called an *inner state transition*. For example, **Figure 11** shows a state transition $S_i \rightarrow S_j$ for a regulator application performing the following control:

$$(y_i, u_{i,j}) \rightarrow (y_j, u_{j,k})$$

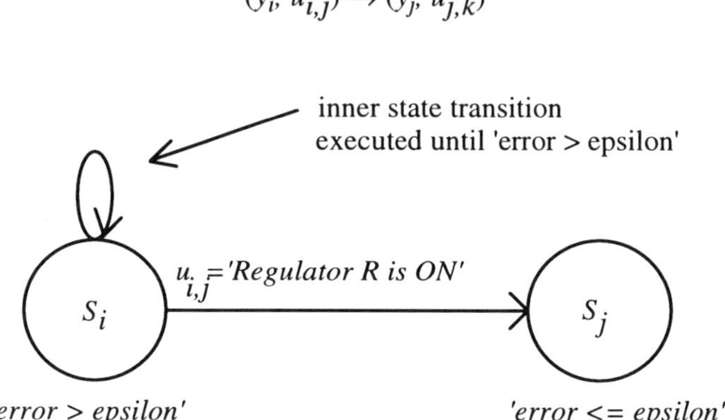

Figure 11 Inner State Transitions.

At time t the state S_i materializes and the controller executes the control action $u_{i,j}$="*regulator **R** is ON*". The plant formula of S_i is y_i="*error>epsilon*" and the plant formula of the consecutive state S_j is y_j="*error $\leq$ epsilon*". In this case, the same state S_i materializes as long as the control action does not bring the error below 'epsilon' .

For the case of inner state transitions, the repeated state transition from the state S_i to itself may occur and thus, S_i is materialized during a number of consecutive sampling instants. For example, if S_i materializes at time t', then it continues to materialize during the next $(k-1)$ sampling instants, before its consecutive state S_j materializes at $t'+k$.

In this chapter, to have a unique time reference to all state transitions, inner or not, we will confine the time needed for a state transition to materialize to the interval $[t, t+1]$, where in the case of inner transitions we have that $t=t'+k-1$ and $t+1=t'+k$.

We would like to stress here that a state $(y_i, u_{i,j})$ with an inner state transition is different from a state of the form $(y_i, u_{i,i})$. The former has a post-condition y_j due to the effect of a control action, but this post-condition materializes only after a time delay, while the later does not have a post-condition due to the control action $u_{i,i}$ (the control action $u_{i,i}$ is the null action).

3.3 Control Knowledge Types

In this section we describe the types of control knowledge required to obtain plant formulas, states, and goal paths in the context of any control application.

The description of an object PC in the previous section makes use of three basic formal constructs: plant formulas, control actions, controller states, and goal paths. The choice of these and their suitability for a particular control application is done by considering application-specific knowledge such as laws of physics, properties of materials and sensors, mathematical models of dynamic behavior and/or simulations, heuristics, and specifications.

We identify three types of domain-specific knowledge, corresponding to the above described formal constructs:

- *Ontological knowledge*: This is application-specific knowledge which is used to determine the components of a plant formula (i.e., constrained plant signals) as well as the form of the boolean function which describes a particular logical configuration of these components for the purpose of control (i.e., it is used to design the plant formulas y_i).

- *Operational knowledge*: This is the application-specific knowledge which is used to decide control actions and their pre- and post-conditions, that is, well-defined states (y_i, $u_{i,j}$).

- *Process knowledge*: This is the application-specific knowledge used to determine optimal goal paths, that is, optimal sequences of state transitions.

In what follows we will discuss each one of the above three types of control knowledge in some detail.

3.3.1 Ontological Knowledge

All three types of control knowledge are an abstraction of the reality obtained say, by an experiment, a mathematical model, a simulation, or a set of heuristics. Independently of the formal precision of such an abstraction, there are always certain constraints which specify under what conditions the application of this abstract knowledge to a particular control application is valid.

Ontological knowledge is the abstracted knowledge underlying the choice of plant formulas. These formulas state explicitly which plant outputs are of interest for control and when interpreted, they show which plant outputs are currently present or not. Thus, the ontological knowledge is used to determine the set of plant signals, the set of constrained plant signals, and finally the logical combinations (boolean functions) of constrained plant signals. Part of the assumptions underlying the validity of the ontological knowledge when used for a particular control application may be explicitly reflected in the plant formulas, but there are normally some assumptions which are implicit in these formulas.

For example, Ohm's law is expressed as the physical formula $R = U/I$. This law can be used in an object PC for the indirect measurement of a current I, using the measured value of U, in which particular case $I = U/R$ becomes a plant formula, i.e., a single plant constraint on the values of the plant signal I. However, the validity of the indirect measurement of I is critically dependent on the resistor R being constant relative to factors such as temperature, magnetic fields, pressure, chemical agents, etc. This dependency is not stated explicitly in the plant formula, and it may be not reflected in say, a specification for the object controller using this plant formula. Thus, the constraint that R should be constant relative to factors such as temperature, magnetic fields, pressure, and chemical agents, is an implicit assumption which when violated in the context of a particular control application would make the above plant formula invalid for this particular control application. This type of unstated assumptions are called *ontological assumptions* since they constitute the basis for the validity of the knowledge (e.g., Ohm's law) used in the construction of plant formulas, i.e., the ontological knowledge. When the ontological assumptions are violated, a plant formula may not materialize since it expresses a condition that cannot be physically realized. If for example, such a plant formula belongs to the state S', a state transition S→S' that is specified to occur cannot materialize.

One may argue here that an ontological assumption can be made a part of a plant formula and thus, become an explicit part of it. In our example, we could have that ($(I = U/R)$ $\wedge(t1 \leq temp \leq t2)$ $\wedge$ $(p1 \leq press \leq p2)$ $\wedge$ *(magn-field-constraint)* $\wedge$ *(chem-agent-constraint)*). However this is not always possible due lack of sensors, measurement costs, complexity considerations, lack of appropriate mathematical models for indirect measurements. For example, if an object PC is to have full control over all factors that may influence the applicability of Ohm's law, it has to be able to measure continuously the temperature, pressure, magnetic fields, chemical substances in the environment, to make sure that they do not exceed certain prescribed limits. Moreover, the controller requires the knowledge about how the aging of the material from which the resistor is built of, affects the allowable limits for the above mentioned factors.

3.3.2 Operational Knowledge

This is the abstracted knowledge used to define actions, the expected effects of actions (action post-condition) on the plant output, and the conditions (action precondition) under which the effect of an action can be expected to take place. Thus, the operational knowledge is directly relevant for defining states and their corresponding consecutive states.

Let us consider the symbolic expression $u_{i,l}$ denoting a physical control action u. This symbolic expression refers to the action precondition y_i and the action post-condition y_l. Similarly, we can write a number of symbolic expressions all denoting the same physical action u but in the context of different pre- and post-conditions, say $u_{i,m}$, $u_{k,r}$ $u_{m,l}$, etc. Purely symbolically all these control actions are equally good candidates for being represented in the object controller and used by it in achieving certain goal states. In this case one decides which symbolic expressions to be represented in the controller by using the operational knowledge available. According to this knowledge, $u_{i,l}$ is deemed as a feasible candidate since only in the case of $u_{i,l}$ one may have that when the precondition y_i is interpreted as true at t, i.e., $y_i(t) = 1$, and the action is executed at t, i.e., $u_{i,l}(t) = 1$, the post-condition y_l will be expected to materialize at $t+1$, i.e., $y_l(t+1) = 1$.

Now let $u_{i,l}$ be executed at t, i.e., $u_{i,l}(t) = 1$, and let also $y_l(t+1) = 1$. From a purely symbolic point of view we have a number of control actions which have y_l as their precondition, namely each one of $u_{l,r}$, $r = 1 ,..., M$. Thus, the controller state $(y_i, u_{i,l})$ has M consecutive states, i.e., each one of $(y_l , u_{l,r})$ is a consecutive state. Here again, the operational knowledge is used to decide which of these M consecutive states are feasible in the sense of the previous paragraph. Consequently, only these states will be represented in the object controller and will form feasible state transitions.

The operational knowledge is used identically to define expected external actions. Since the exact time of the occurrence of such an action is not known in advance its precondition may in principle be any plant formula. The operational knowledge is then used to determine for exactly which plant formulas the execution of an external action can change the plant output and in what manner. Thus, only a subset of all plant formulas is used to describe the pre- and post-conditions of the set of external actions.

Thus, it is the operational knowledge which ensures that preconditions, actions, and post-conditions are linked in certain way, so that a controller state is in principle feasible to materialize. This in turn ensures the materialization of a consecutive state which has the expected effect of an action as the precondition for another action.

For example, in a power generation plant with two generators (see example 2 in Chapter 2), the load controller starts a second generator if the load on the first generator is too high (above some maximum admissible load) with the immediate purpose to reduce the load on the first generator. Let this *start-generator* control action be executed when the load on a generator is less or equal the maximum admissible load, but greater or equal some minimal admissible load. Then the effect of this control action will be that the load on both generators will become less than the minimum admissible load which is undesirable.

In this context, the control action *start-generator-2* , say $u_{i,l}$, has as its precondition the plant formula $y_i = L1 > Lmax$ and as its post-condition the plant formula $y_l = (Lmin < L_1 < Lmax) \wedge (Lmin < L_2 < Lmax)$, where L_1 and L_2 are the loads on generator-1 and generator-2 respectively; *Lmax* is the maximum load on one generator; *Lmin* is the minimum load on one generator. The plant formula y_i represents a plant output such that the load on generator-1 is higher than the maximum admissible load. If this plant output really takes place at t, i.e., $y_i(t) = 1$, the control action *start-generator-2* is executed at t and it is expected that at $t+1$ the expected effect (the post-condition) y_l of this control action will take place. That is, the load on both generators will become less than the maximum admissible load since at $t+1$ it will be shared by both generators.

Let us observe here that the control action *start-generator-2* can be executed at any time even when its precondition $y_i = L1 > Lmax$ does not interpret as true. However, in such a case the post-condition of the control action would not materialize since, the load on both generators will then become less than the minimum admissible load which is an undesirable situation.

Let the control action start-generator be physically executed by another, lower level controller, say a diesel engine controller. The operational knowledge is used to define the control action *start-generator-2* of the load controller, and to relate this action to particular pre- and post-conditions. This operational knowledge is the control knowledge (ontological, operational and process) used in the design of the diesel engine controller such as: how is the speed of the diesel engine measured; thermodynamical properties of the diesel engine; how to start the diesel motor; how to synchronize the electric generators; how to connect a breaker; how to balance the load.

Here again there is a set of assumptions underlying the control knowledge used in the design of the diesel engine controller and thus the operational knowledge for the load controller. When these assumptions are wrong, or correct but violated during control, the control actions executed by the load controller will not realize their expected effects. One can see here that defining the assumptions underlying the operational knowledge about control actions is of a recursive nature: the assumptions at a certain control level are defined as the assumptions underlying the control knowledge used in the design of a lower control level.

The complexity of the operational knowledge explains why control with distributed hierarchical PCs is so sensitive to a large number of assumptions underlying its correctness per se as well as its applicability to a particular control application.

3.3.3 Process Knowledge

Normally, there is a set of feasible state transitions from an initial state to a goal state which determine a set of goal paths. However, only one of these goal paths is optimal in some control theoretic sense (i.e. intended for the application).

For example, let us consider the process knowledge for the production of four chemical substances in a chemical plant described in **Figure 12**.

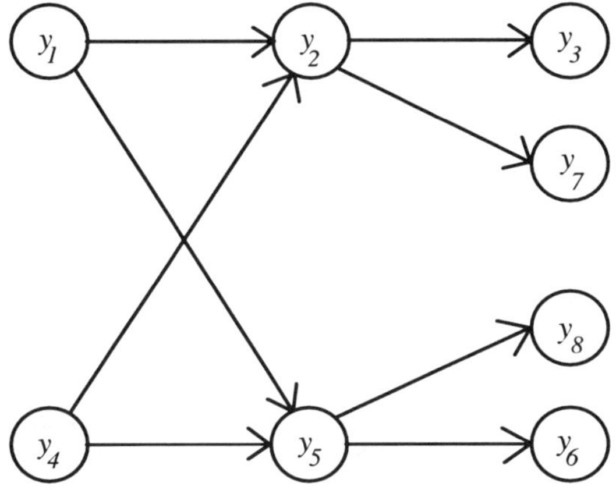

Figure 12 Example of process knowledge for a chemical plant.

Each plant formula y_i ($i=1, \ldots ,8$) represents a plant output in terms of a chemical substance. Then $y_i(t) = 1$ means that a particular chemical substance is present at t. Furthermore, when a chemical substance is present, it is used in the production of another chemical substance. For example, y_2 can be used in the production of y_3 or y_7. The goal of the production process is the realization of any one from the chemical substances y_3, y_7, y_6, and y_8. Let all physically feasible goal paths from the initial plant formulas y_1 and y_4 to the plant formulas y_3, y_7, y_6, y_8 which are parts of goal states, be the following:

- $(y_1, u_{1,2}) \rightarrow (y_2, u_{2,3}) \rightarrow (y_3, u_{3,3})$

- $(y_1, u_{1,2}) \rightarrow (y_2, u_{2,7}) \rightarrow (y_7, u_{7,7})$

- $(y_1, u_{1,5}) \rightarrow (y_5, u_{5,6}) \rightarrow (y_6, u_{6,6})$

- $(y_1, u_{1,5}) \rightarrow (y_5, u_{5,8}) \rightarrow (y_8, u_{8,8})$

- $(y_4, u_{4,2}) \rightarrow (y_2, u_{2,3}) \rightarrow (y_3, u_{3,3})$

- $(y_4, u_{4,2}) \rightarrow (y_2, u_{2,7}) \rightarrow (y_7, u_{7,7})$

- $(y_4, u_{4,5}) \rightarrow (y_5, u_{5,6}) \rightarrow (y_6, u_{6,6})$

- $(y_4, u_{4,5}) \rightarrow (y_5, u_{5,8}) \rightarrow (y_8, u_{8,8})$

However, according to the manufacturing specification and the nature of the chemical processes involved, the state transitions which provide the achievement of the goal chemical substances in an optimal manner are only the following:

- $(y_1, u_{1,2}) \rightarrow (y_2, u_{2,3}) \rightarrow (y_3, u_{3,3})$

- $(y_1, u_{1,5}) \rightarrow (y_5, u_{5,8}) \rightarrow (y_8, u_{8,8})$

- $(y_4, u_{4,2}) \rightarrow (y_2, u_{2,7}) \rightarrow (y_7, u_{7,7})$

- $(y_4, u_{4,5}) \rightarrow (y_5, u_{5,6}) \rightarrow (y_6, u_{6,6})$

The operational knowledge is used (via the use of actions and their pre- and post-conditions) to determine the first and second set of state transition sequences but it may not be used to restrict the first set of state transition sequences to the second one. It is the process knowledge which is used to decide why certain sequence of state transitions is optimal while others are not.

3.4 Control With An Object PC

In this section we describe how control with a PC is realized on a given set of goal paths by the use of the interpretation, state completion, goal seeking, and synchronization operations. First we describe the interpretation and the completion operations. Second, we present in a formal manner the goal seeking operation and identify this part of it which corresponds to the synchronization operation. Then we prove the formal properties of the goal seeking operation.

3.4.1 The Interpretation Operation

This operation is used by the object PC to interpret plant formulas given certain measured/estimated values of plant signals and to interpret (execute) control actions. Thus, when both the plant formula and the control action of a state are interpreted to 1, the interpretation operation materializes a state. However, when only a plant formula y_i is interpreted to true, a partial image state $(y_i, -)$ is materialized by the interpretation operation. When a control action is interpreted to true at t we will also say that this control action is executed at t. When a plant formula is interpreted to true at t we will also say that this plant formula is materialized at t.

3.4.2 The State Completion Operation

This operation is used by the object controller to complete a partial material image state with a control action so that the completed state is well defined.

In general, the completion operation is part of the goal seeking operation. This is the case when one and the same plant formula appear in more than one state and thus, the control action from any one of these states can be used to complete the given partial material state. Since these states may be on different goal paths, the goal seeking operation is needed to determine a control action, so that when the partial material state is completed with this control action it will be a state on the goal path with the highest priority. However, when each plant formula belongs to only one state, then there is a unique control action associated with each plant formula. Thus, the control action which can be used to complete a given partial material state is uniquely determined and there is no need to evoke the goal seeking operation.

However, even in this latter case, there is still a need for the goal seeking operation since in a problematic control situation the currently interpreted goal path is aborted and the interpretation has to continue from a partial material state on a pending goal path and there may be more than one pending goal paths.

3.4.3 The Control Scheme

At any sampling instant t the object PC "reads" the values, $x_i(t)$, of plant signals, x_i, and then uses these values to interpret the elements of the set of plant formulas Y. The details of the interpretation of plant formulas were already described in Section 2.2.1.

Let the goal state pursued be G. Suppose that at t-2 the control action $u_{i,j}$ is executed (or interpreted to true) given that $y_i(t$-$2)=1$. Let the plant formula obtained by the interpretation operation at t-1 be y_j, i.e., $y_j(t$-$1) = 1$. Thus, at t-1 we obtain a partial image state, i.e., $(y_j, -)$. Its control action remains to be identified with a particular control action, so that this partial state becomes a well-defined state on a goal path to G. If there is a unique state say, $(y_j, u_{j,n})$, in the state set of the object PC, then the completion operation completes the partial state $(y_j, -)$ with the control action from this unique state. Then the interpretation operation interprets $u_{j,n}$ to true.

As already described, a particular plant formula may be a precondition for a number of different control actions. In our example, y_j may be a precondition for each one of $u_{j,k}$ where $k \in \{ 1,2, \ldots ,M \}$. The states $(y_i, u_{i,k})$ are called *candidate states relative to* y_i. Some of the candidate states may be on a goal path leading to a goal that is different from G.

In this case, the object controller uses the goal-seeking operation to determine a unique control action from the set $\{u_{j,k}\}$, $k \in \{1,2, \ldots ,M \}$. According to the definition of a well-defined state only an element of $\{u_{j,k}\}$ can be used, so that the partial state $(y_j, -)$ becomes a well-defined state. The control action (an element of $\{u_{j,k}\}$, $k \in \{ 1,2, \ldots ,M \}$) is chosen by the goal seeking operation in such a way that the obtained well-defined state belongs either to the optimal goal path to G, or belongs to a non-optimal goal path to G.

When such a control action, say $u_{j,r}$, is determined, the interpretation operation interprets $u_{j,r}$ to true i.e., $u_{j,r}(t$-$1)=1$. Thus, the material state of the object controller at t-1 becomes $(y_j(t$-$1), u_{j,r}(t$-$1))$. Once this material state is obtained, the plant formula expected to interpret to true at the next sampling instant t is y_r.

At t, the object PC samples again the plant signals, x_i, and the interpretation operation determines which plant formula interprets to true. If the object PC performs according to expectations, then we will have that $y_r(t) = 1$, that is, the actual plant output is exactly the one expected as a result of executing the control action $u_{j,r}$ at t-1. Let it be the case that $y_r(t) = 1$, or we have the partial state $(y_r, -)$. The completion operation is applied again to determine a control action having y_r as its precondition. Let this control action be $u_{r,k}$. Consequently, another well-defined state, $(y_r, u_{r,k})$, is found, such that it is: (*i*) a consecutive state of $(y_j, u_{j,r})$ (e.g., the controller state at t-1), and (*ii*) belongs either to the optimal goal path to G or to a non-optimal goal path to G. Thus, we have so far the states $(y_j, u_{j,r})$ and $(y_r, u_{r,k})$

which are consecutive, and both of them are on the optimal goal path to G, or on a non-optimal goal path to G.

The control action $u_{r,k}$ is executed at t and at $t+1$ the sequence of interpretation and completion operations is performed again and another well-defined state (the consecutive state of $(y_r, u_{r,k})$) on the optimal goal path to G or on a non-optimal goal path to G is determined.

The above described control scheme is repeated until the goal state is reached, i.e., the plant formula component of G is interpreted to true. The achievement of the goal state G is guaranteed by the goal seeking operation since it is required to produce completed states that are either on the optimal goal path to this goal state, or on a non-optimal goal path to it. Furthermore, as long as the expected result of each control action is interpreted as true, the goal-seeking operation guarantees that the object controller follows all the time the optimal goal path to G.

We would like to stress here again that when the following two restrictions hold:

- each y_i appears only in one plant formula and thus, each y_i is associated with a unique control action $u_{i,j}$, and

- the expected result y_j of $u_{i,j}$ always materializes (i.e., interprets to true)

then the GSO is not needed to assist the completion operation in the completion of a partial image state. The reason for this is as follows. Since each y_i has a unique control action $u_{i,j}$ associated with it this implies that each state can only be in one optimal goal path. Furthermore, we will also have that the state $(y_i, u_{i,j})$ will have only one consecutive state say $(y_j, u_{j,n})$ on the same optimal goal path. Thus, when the state $(y_i, u_{i,j})$ is materialized at t and the expected effect y_i of $u_{i,j}$ always is materialized at $t+1$ there will be only one control action which can complete the partial state $(y_j, -)$, namely $u_{j,n}$. Hence, once the interpretation operation has started to interpret a particular goal path the completion operation will always complete partial image states in such a way so that they are on the same optimal goal path. However, the second of the above two restrictions is hardly a realistic one and in practice the completion operation is in fact part of the goal seeking operation.

In the next section we will describe in detail the goal seeking operation (GSO).

3.4.4 The Goal Seeking Operation

In this section we describe in detail the goal seeking operation of an object PC and the properties of this operation. However, the particular implementation of the goal-seeking operation is not relevant for the purpose of ontological control. What is relevant for ontological control is the type of control knowledge the GSO uses and how the GSO uses it to complete a partial image state such that it achieves the current goal state being pursued. In what follows we describe the goal seeking operation from this perspective. First we describe the type of control knowledge used by the goal seeking operation in terms of the so called extended image states. Then we describe the operation of the GSO, and use this description to prove the properties of the GSO relevant for ontological control.

The goal-seeking operation described here uses the process and operational types of control knowledge. As already described, the process knowledge determines a state transition sequence corresponding to an optimal goal path. The operational knowledge determines well-defined states. Furthermore, each well-defined state $(y_i, u_{i,j})$ on an optimal goal path is called an *image state*, and one and the same image state as well as one and the same partial state can be common for a number of optimal goal paths.

Control Knowledge For The GSO

As already mentioned in the previous section, the purpose of the GSO is to complete a partial image state $(y_i, -)$ with such a control action $u_{i,j}$ so that the present goal state G_k being pursued is reachable from the completed state via an optimal / non-optimal goal path.

The control knowledge required for the achievement of the above stated purpose and associated with each image state $S = (y_i, u_{i,j})$ is as follows:

- The set of goal states reachable from $(y_i, u_{i,j})$ via optimal goal paths. This set is denoted as OGS and since each state can be on at least one optimal goal path it has at least one element;

- Each image state from OGS has the so called state-index associated with it. The initial state on an optimal goal path has the state-index *1*. The consecutive state of the initial state on this same optimal goal path has the state-index 2, etc. Thus, if $(y_i, u_{i,j})$ belongs to the optimal goal path to G, and there are *m-1* image states before it on the this optimal goal path the state-index of $(y_i, u_{i,j})$ will be *m*. Observe here, that since $(y_i, u_{i,j})$ can also be on the optimal goal path to a different goal state G' it may have a different state-index on the optimal goal path to this other goal state.

- The set of goal states reachable from $(y_i, u_{i,j})$ via non-optimal goal paths. This set is denoted as NOGS. Recall here (Section 3.2.2) that one and the same goal state G can be achieved both via an optimal goal path and at least one non-optimal goal path. Since one and the same image state can be on more than one different optimal goal paths it can have, for example G in both of OGS and NOGS.

- The priority order between goals. Every two goal states G and G' are related to each other by a priority relation, $>$, reading as *the priority of G is greater than the priority of G'*. This relation is reflexive, antisymmetric, and linear. This in turn implies that there is a strict linear order on the set of goal states. That is, for each two goal states G and G' we have that either $G > G'$ or $G' > G$, and for each subset of the set of goal states there is a unique goal state with minimum priority and a unique goal state with maximum priority. The priority order on goal states defines the order in which the goal states have to be achieved: $G > G'$ means that G' can be pursued iff G is achieved or its achievement is impossible.

- Present goal state (PGS). This is the present goal state being pursued. Suppose that the priority order on goal states is $G_1 > G_2 > ... > G_k > G_{k+1} > ... > G_n$. If the present goal state is, say G_k this means that each one of G_1, G_2, ... G_{k-1}, has either been achieved or its achievement has failed, and G_k has at the present moment the highest priority over G_{k+1}, ... , G_n.

All this additional information is used to define an *extended image state* which is used by the GSO. Thus, if $(y_i, u_{i,j})$ is an image state, its extended version will be as follows:

$$<(y_i, u_{i,j}); \text{PGS} ; \text{OGS} ; \text{NOGS} ; '>' >$$

where each element of OGS is a tuple of a goal state which is reachable from $(y_i, u_{i,j})$ via an optimal goal path and the state index of $(y_i, u_{i,j})$ on this optimal goal path. In the above five-tuple its last element '>' is a notation for the priority order between goal states. That is, $>$ stays for say, $G_1 > G_2 > ... > G_k > G_{k+1} > ... > G_n$. Furthermore, let us note here that for each image state $(y_i, u_{i,j})$ we have that:

- its OGS set is never empty since each image state $(y_i, u_{i,j})$ is on at least one optimal goal path;

- its NOGS set may be empty or contain at least one element. Furthermore, the intersection of OGS and NOGS may be empty or not.

- the priority order is established during design and does not change during control.

Finally, let us observe here that the optimal and non-optimal goal paths are implicit in the set of extended image states. For example, each image state $(y_i, u_{i,j})$ on the optimal goal path to G_k will have this goal state in its OGS set. Furthermore, the state index associated with G_k in this set indicates the exact position of $(y_i, u_{i,j})$ on the optimal goal path to this goal state. In a similar manner, each state on a non-optimal goal path to G_k will have this goal state in its NOGS set.

The Goal Seeking Operation

Let the present goal state be G_k. Thus, we have that this goal state has the highest priority amongst G_{k+1} , ... , G_n and each one of G_1 , ... , G_{k-1} has either been already achieved or its achievement has failed. Let at t the material image state be $(y_k, u_{k,i})$, i.e., $y_k(t) = 1$ and $u_{k,i}(t) = 1$. Thus, since $u_{k,i}$ is executed at t, its post-condition y_i is expected to materialize at $t+1$. Suppose that at $t+1$ we have that $y_i(t+1) = 1$. Thus, we obtain the partial image state $(y_i, -)$. Now the GSO has to complete this partial state with a control action so that the completed state is well-defined. This implies that the control action has to be chosen from the control action parts of the consecutive image states of $(y_k, u_{k,i})$.

Recall here that for each image state its set of consecutive states is constant. The GSO proceeds through the following steps (an overview is shown in **Figure 13**):

STEP 0

The GSO checks first whether $(y_i, -)$ matches the plant formula part of G_k. If there is a match, then G_k is achieved. The GSO completes the partial state $(y_i, -)$ with the action $u_{i,i}$, i.e. the goal state is $(y_i, u_{i,i})$. The present goal state becomes G_{k+1} that is, the goal state with priority immediately below this of G_k in the priority order over goal states. Note that this step performs the completion operation followed by an update of the currently pursued goal path. Then the controller starts from the initial state on the optimal goal path to G_{k+1}, materializing it and returning to STEP 0 once the result of the execution of the control action from this initial state is obtained. If there is no match between $(y_i, -)$ and G_k the GSO goes to STEP 1.

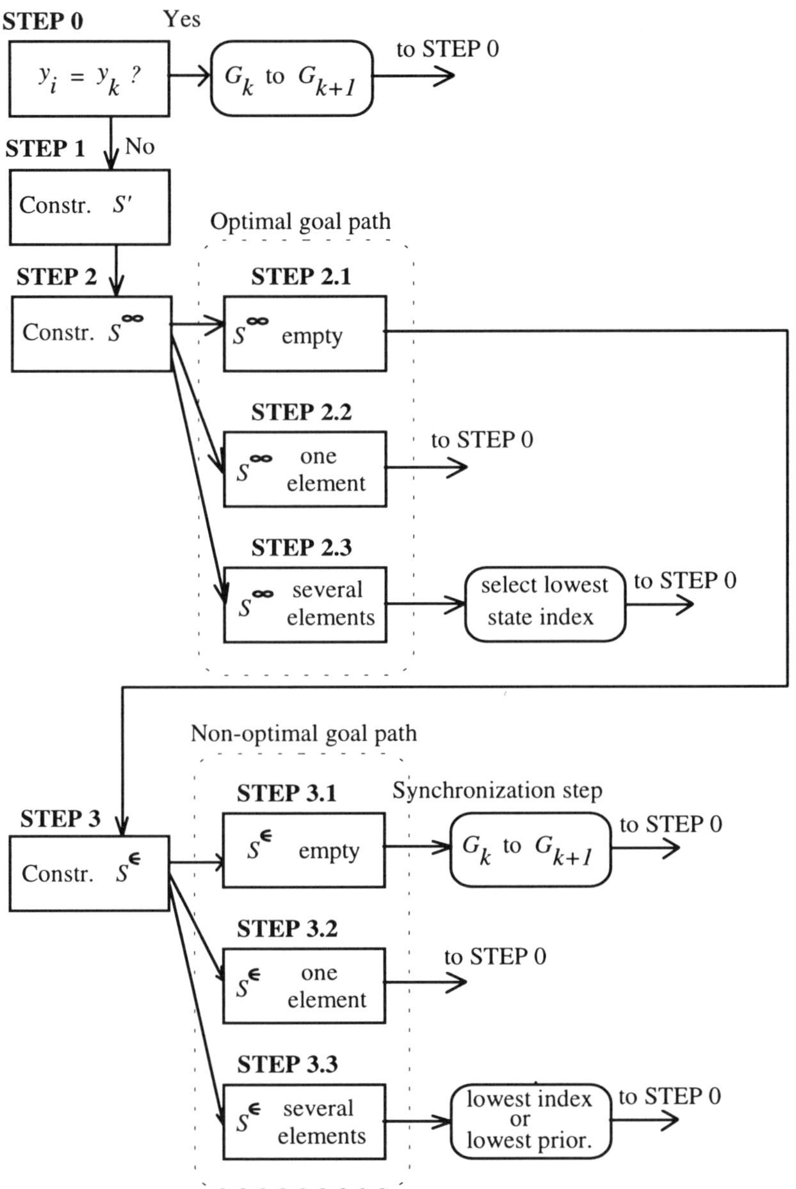

Figure 13 GSO Overview.

STEP 1

At this step the GSO has to construct a set S' of image states such that an image state belongs to S' iff it has y_i in its plant formula part. In other words, the elements of S' are the consecutive states of $(y_k, u_{k,i})$. Since each image state different from a goal state has at least one consecutive state, S' cannot be empty. Once S' is constructed the GSO goes to STEP 2.STEP 2

The GSO constructs the set S^∞ which is a subset of S' and an image state belongs to it iff its OGS set contains the present goal state G_k. In other words, G_k is reachable via an optimal goal path from each element in S^∞. Depending on whether S^∞ is empty, contains a single element, or contains more than one elements, the GSO goes to STEP 2.1, STEP 2.2, and STEP 2.3 respectively.

STEP 2.1

Let S^∞ be empty. This implies that none of the image states in S' is on the optimal goal path to G_k. The GSO goes to STEP 3.

STEP 2.2

Let S^∞ have a single element, say $(y_i, u_{i,r})$. In other words, there is only one image state $(y_i, u_{i,r})$ in S' which is on the optimal goal path to G_k, and it is the only consecutive image state of $(y_k, u_{k,i})$. The GSO selects $(y_i, u_{i,r})$ as the completed version of $(y_i, -)$ and materializes this state at $t+1$ and at $t+2$, when a new partial state is obtained as the result of executing $u_{i,r}$, it starts again from STEP 0. Note that this step performs the state completion operation.

STEP 2.3

Let S^∞ have more than one elements. In other words, there is more than one image states in S' that is on the optimal goal path to G_k. In this case the GSO selects the image state which has the lowest state-index, say $(y_i, u_{i,s})$, as the completed version of $(y_i, -)$. The completed image state $(y_i, u_{i,s})$ is materialized at $t+1$ and at $t+2$ the GSO starts from STEP 0 trying to complete the partial image state obtained as the result of the execution of $u_{i,s}$ at $t+1$. (The lowest state-index is selected due to safety reasons: the control algorithm should be executed fully).

STEP 3

At this step the GSO has to construct a set $S^\in$ that is a subset of S' such that an image state belongs to it iff its NOGS set contains G_k. In other words, since there is no image state (empty S^∞) from which G_k is reachable via an optimal goal path, the GSO tries to find image states from which G_k is reachable via a non-optimal goal path. Here again, depending on whether $S^\in$ is empty, contains a single element, or contains more than one elements, the GSO goes to STEP 3.1, STEP 3.2, and STEP 3.3 respectively.

STEP 3.1

Let $S^\in$ be empty. This implies that there is no consecutive image state of $(y_k, u_{k,i})$ which is on a non-optimal goal path to G_k. At the same time none of the consecutive states of $(y_k, u_{k,i})$ is on the optimal goal path to this same goal state. These two facts imply that G_k is simply not reachable from $(y_k, u_{k, i})$. At this stage, the object PC has to determine a new optimal goal path. This corresponds to the controller being *de-synchronized*, as already described in Chapter 2.

In this case the GSO performs the so called *synchronization step* as follows: the GSO replaces the present goal state G_k with the goal state whose priority is immediately below that of G_k in the priority order. Thus, the new present goal state becomes G_{k+1} and the GSO goes to STEP 0 trying to complete $(y_i, -)$ given that the present goal state is G_{k+1}. If the synchronization step fails for the goal state G_{k+1} it is repeated for the goal state whose priority is immediately below that of G_{k+1}, etc., until $(y_i, -)$ is completed (the completion is guaranteed since the state set S is ontologically complete). It can be easily seen that the *synchronization operation* from Chapter 2 is described here as a recursive interaction between the *synchronization step* of the GSO and the rest of the GSO operation.

STEP 3.2

Let $S^\in$ have a single element, say $(y_i, u_{i,t})$. In other words, there is only one image state $(y_i, u_{i,t})$ in S' which is on a non-optimal goal path to G_k and it is the only consecutive image state of $(y_k, u_{k,i})$. The GSO selects $(y_i, u_{i,t})$ as the completed version of $(y_i, -)$ and materializes this state at $t+1$. Thus, at $t+2$, when a new partial state is obtained as the result of executing $u_{i,r}$, the GSO starts again from STEP 0. Note that this step performs the state completion operation.

STEP 3.3

Let $S^\in$ have more than one element. In other words, there is more than one image state in S' which is on non-optimal goal paths to G_k. These image states may be all on the same non-optimal goal path to G_k (if there is only one non-optimal goal path to G_k), or they may belong to different non-optimal goal paths to G_k.

In the case when all image states in $S^\in$ belong to the same non-optimal goal path, the GSO selects the image state which has the lowest state-index, say $(y_i, u_{i,q})$, as the completed version of $(y_i, -)$. The completed image state $(y_i, u_{i,q})$ is materialized at $t+1$ and at $t+2$ the GSO starts from STEP 0 trying to complete the partial image state obtained as the result of the execution of $u_{i,q}$ at $t+1$.

In the case when the elements of $S^\in$ belong to different non-optimal goal paths the image states which have G_{k+1} in their OGS sets are selected. If there are no such image states then the image states which have G_{k+2} in their OGS sets are selected, etc. Thus, the so selected image states will all belong to one and the same non-optimal goal path to G_k. From these image states the GSO selects further the image state with the lowest state index, say $(y_i, u_{i,m})$. The completed image state $(y_i, u_{i,m})$ is materialized at $t+1$ and at $t+2$ the GSO starts from STEP 0 trying to complete the partial image state obtained as the result of the execution of $u_{i,m}$ at $t+1$.

The Properties Of The GSO

The goal seeking operation described in the previous section has two important properties:

- If the GSO starts from an image state on the optimal goal path to the present goal state G_k and the expected post-condition of each control action is always materialized then the GSO will achieve G_k via the optimal goal path to this goal state.

- If the GSO starts from an image state on a non-optimal goal path to G_k, and the expected post-condition of each control action is always materialized then the GSO will achieve G_k by following one and the same non-optimal goal path to this goal state.

In what follows we will prove the above two properties of the GSO.

Proposition 3:1 *Let $S \to S'$ be an arbitrary state transition on the optimal goal path to the present goal state G_k and let the image states S and S' be defined as $S = (y_i, u_{i,j})$ and $S' = (y_j, u_{j,k})$. If the control action $u_{i,j}$ is executed at t, given that $y_i(t) = 1$, and its expected result y_j interprets as true at t+1, then the partial state $(y_j, -)$ will be completed by the GSO in such a way so that it is equivalent with S'.*

Proof.

Let the state-indices of S and S' be l and $l+1$ respectively. Recall also that there is a unique image state corresponding to a particular state index within the same optimal goal path. Furthermore, let us observe here that the OGS set of S' contains the present goal state G_k since S' is on the optimal goal path to G_k. However, since S' can be on different optimal goal paths its OGS set may contain other goal states as well.

Since the expected post-condition y_j of $u_{i,j}$ is interpreted as true at $t+1$ the input to STEP 0 of the GSO will be the partial state $(y_j, -)$. Let us now consider the behavior of the GSO step by step.

Step 0:

If y_j is in the plant formula part of the G_k and since y_j is interpreted as true at $t+1$ this implies that G_k is achieved. The GSO completes the partial state $(y_j , -)$ with the action $u_{j,j}$, i.e. the goal state is $(y_j, u_{j,j})$. In this case the GSO will choose a new goal state and continue from the initial image state on the optimal goal path to this new goal state.

If y_j is not in the plant formula part of the G_k the GSO goes to STEP 1 looking for ways to complete $(y_j, -)$.

Step 1:

The set S' containing image states which all have y_j in their plant formula parts is not empty since the set of consecutive states of $S = (y_i , u_{i,j})$ is not empty. This follows from the fact that S is part of the state transition $S \to S'$ on the optimal goal path to G_k and thus, at least S' is in the set of its consecutive states.

Step 2.1:

The set S^{∞} containing only these image states from S' which have G_k in their respective OGS sets is not empty. This follows from the fact that S' is in S' and it has G_k in its OGS set. Thus, there is at least one image state in S^{∞}, namely $S' = (y_j$, $u_{j,k})$. If there are other than S' image states in S^{∞} they will have y_j in their plant formula parts, but different from $u_{j,k}$ control actions in their control action parts. Furthermore, they will all have G_k in their OGS sets.

Step 2.2:

If S^{∞} has a single image state than this image state is equivalent to $S' = (y_j, u_{j,k})$. Suppose that it is instead $S'' = (y_j , u_{j,r})$ and $S' \neq S''$, i.e., $u_{j,r} \neq u_{j,k}$. Thus, there is a state transition $S \rightarrow S''$ on the optimal goal path to G_k originating at S. Consequently, there are two state transitions, $S \rightarrow S'$ and $S \rightarrow S''$, originating from the same image state S and both on the optimal goal path to G_k. Since, for every goal path there is a unique transition originating at an image state and ending with its consecutive state on the same goal path, the above situation is impossible. Thus, it can only be the case that $S' \equiv S''$.
Q.E.D.

Step 2.3:

Here S^{∞} contains not only S' but a number of other image states which all have y_j in their plant formula parts and G_k is in their OGS sets. The latter implies that all these image states are on the optimal goal path to G_k. Thus, each image state in S^{∞} will have a state-index different from the state-index $l+1$ of $S' = (y_j , u_{j,k})$. Suppose for the time being that all these state-indices are greater than $l+1$. Since at this step, the GSO will choose the image state with the lowest state-index, this state will be the one with state-index $l+1$. Consequently, since to each state-index corresponds a unique state, within one and the same optimal goal path, the chosen state will be $S' = (y_j , u_{j,k})$.
Q.E.D.

Now suppose, without loss of generality, that S^{∞} contains not only S' (state-index $l+1$), but also an image state, $S'' = (y_j , u_{j,r})$, in S^{∞}. Let this image state have a state-index $l-n < l+1$. Furthermore, let there be no other image state in S^{∞} whose state index k is such that $l-n < k < l+1$ and $k < l-n$. Thus, S'' is closer to the initial state of the optimal goal path to G_k than S' and it is the only consecutive state of S' which has a lower state-index than S'.

In this case, the GSO, as already described, will choose the image state S'' as the completed version of $(y_j, -)$ instead of $S' = (y_j, u_{j,k})$. Once S'' is materialized the GSO will proceed further from STEP 0 and will materialize the image states on the goal path to G_k with state-indices l-n, l-$n+1$, ... ,l, where the latter state-index corresponds to S. When S is materialized, it will be again the partial state $(y_j, -)$ which is to be completed at the next sampling instant. Since, S^∞ at this stage contains at least S' and S'' (other image states in this set will have state-indices greater than $l+1$) it will again choose S'' because l-$n < l+1$. Thus the sequence of image states, all on the optimal goal path to G_k, and with state-indices l-n, l-n $+1$, ..., l will be repeated an indefinite number of times and as a consequence, S' will never materialize.

The above obviously defies Proposition 3:1 in the particular case when S^∞ contains not only S' but also image states with state-indices that are lower than this of S'.

However, the presence of cycles (repetitive occurrence of one and the same state transition sequence) is not an undesired feature: such cyclic behavior of the GSO is sometimes required because of control considerations. In such cases, the GSO is augmented with additional knowledge so that after a state transition sequence is repeatedly executed a pre-specified number of times (0, 1, 2, ...), the GSO will delete S'' from the list of elements in S^∞ and thus, only the consecutive states of S with state-indices higher than l will remain in S^∞. Thus, as already illustrated in STEP 2.2, the state S' will be chosen as the completed version of $(y_j,-)$ since every other image state in S^∞ will have a higher state-index.
Q.E.D.

The type of knowledge required so that the GSO can leave a cycle after having repeated it a pre-specified a number of times is not relevant for ontological control. What is important, is that the GSO can recognize such "normal" cycles and then is able to execute them as required. Since Proposition 3:1 holds for any state transition on the optimal goal path to G_k this guarantees that having first materialized the initial state on this optimal goal path (state-index 1), the GSO will then choose the image state with state-index 2 on this same optimal goal path, etc. till G_k is finally achieved. This will of course be the case as long as the expected post-condition of each control action being executed materializes at the next sampling instant.

Proposition 3:2 *Let $S = (y_i, u_{i,j})$ not belong to the optimal goal path to the present goal state G_k, but belong to at least one non-optimal goal path to this same goal state. If the GSO starts from S and the expected postconditions of all control actions materialize, then the GSO will reach G_k by following one and the same non-optimal goal path to it.*

Proof.

Recall first that the existence of a non-optimal goal path to G_k implies the following:

- Every image state on a non-optimal goal path to G_k will have G_k in its NOGS set;

- Every image state which is both on the optimal goal path to G_k and on a non-optimal goal path to this same goal state will have G_k in its OGS set and in its NOGS set;

- Every image state which is not on the optimal goal path to G_k will not have this goal state in its OGS set but will have G_k in its NOGS set.

- Each image state belongs to at least one optimal goal path;

- For each image state its set of consecutive states is constant.

Let at t the image state be $S = (y_i, u_{i,j})$ such that G_k does not belong to its OGS set, but belongs to its NOGS set. When $u_{i,j}$ is executed at t, given that $y_i(t) = 1$, its expected post-condition, y_j is materialized at $t+1$, and thus the partial state $(y_j, -)$ is obtained and has to be completed by the GSO. Let us now again follow the behavior of the GSO through its different steps.

STEP 0

If y_j is in the plant formula part of the G_k and since y_j is interpreted as true at $t+1$ this implies that G_k is achieved. The GSO completes the partial state $(y_j, -)$ with the action $u_{j,j}$, i.e. the goal state is $(y_j, u_{j,j})$. In this case the GSO will choose a new goal state and continue from the initial image state on the optimal goal path to this new goal state.

If y_j is not in the plant formula part of the G_k the GSO goes to STEP 1 looking for ways to complete $(y_j, -)$.

STEP 1

The set S' of all image states containing y_j in their plant formula parts is not empty. This follows from the fact that S belongs to at least one optimal goal path and since it is not a goal state, it will have at least one consecutive state on this same optimal goal path.

STEP 2

The set S^∞ of image states which are on the optimal goal path to G_k is to be constructed from the image states in S'.

STEP 2.1

Let S^∞ be empty. This implies that no image state in S' is on the optimal goal path to G_k and the GSO proceeds to STEP 3.

STEP 2.2

Let S^∞ have a single image state, say $S' = (y_j , u_{j,m})$, i.e., S' is on the optimal goal path to G_k and is the only consecutive state of S such that it is on the optimal goal path to G_k. Since the set of consecutive states of S is fixed, this implies that every time the GSO starts from S, and the present goal state is G_k, it will always choose S' as the completed version of $(y_j, \text{-})$. According to Proposition **3**:1, since S' is on the optimal goal path to G_k and the expected post-condition of each control action does always materialize, the partial state $(y_m, \text{-})$, obtained after the materialization of S', will be the consecutive state of S' on the optimal goal path to G_k, etc. Thus, starting from S, the GSO will always achieve G_k, by choosing one and the same state transition sequence, $S \rightarrow S' \rightarrow ... \rightarrow G_k$.
Q.E.D.

STEP 2.3

Let S^∞ have more than one image state in it, i.e., S has a number of image states all these image states are on the optimal goal path to G_k. In this case the GSO chooses the state with lowest state-index, say $S' = (y_j, u_{j,m})$. Since the set of consecutive states of S is fixed and all states in it have different state-indices (since they all are on the optimal goal path to G_k), the chosen image state will always be S' whenever the GSO starts from S. According to Proposition **3**:1, since S' is on the optimal goal path to G_k and the expected post-condition of each control action does always materialize, the partial state $(y_m, -)$, obtained after the materialization of S', will be the consecutive state of S' on the optimal goal path to G_k, etc. Thus, starting from S, the GSO will always achieve G_k, by choosing one and the same state transition sequence, $S \rightarrow S' \rightarrow ... \rightarrow G_k$.

Q.E.D.

STEP 3

Recall here that we arrive at this step when S^∞ is empty, that is no consecutive state of S is on the optimal goal path to G_k.

Furthermore, the GSO tries to construct $S^\in$ such that each image state in it has G_k in its NOGS set.

STEP 3.1

$S^\in$ cannot be empty. The reason for this is that S belongs to at least one optimal goal path and thus, has at least one consecutive image state, namely its consecutive image state on this same optimal goal path. Consequently, this step is skipped and the GSO goes to STEP 3.2

STEP 3.2

Let $S^\in$ have a single element, say $S' = (y_j, u_{j,p})$, i.e., it is a the only consecutive state of S on a non-optimal goal path to G_k. This image state is then chosen as the completed version of $(y_j, -)$, and materialized at $t+1$. Consequently, y_p is interpreted as true at $t+2$ and the GSO has to complete the partial state $(y_p, -)$.

If STEP 2.1 or STEP 2.2 are repeated while trying to complete $(y_p, -)$ then we are guaranteed that the completed state will be on the optimal goal path to G_k, and from there, Proposition **3:1** further guarantees that the GSO will reach this goal state on the optimal goal path to it. Furthermore, since in this particular case S has a unique consecutive state S', whenever the GSO starts from S it will always go via the sequence $S \rightarrow S' \rightarrow ... \rightarrow S_k$.

If instead of STEP 2.1 or STEP 2.2 the GSO goes a number of times to STEP 3.2 this implies that every image state on the non-optimal goal path to G_k, starting from S, has a unique consecutive state, and thus whenever the GSO starts from S the GSO will always repeat one and the same state transition sequence.

Q.E.D.

STEP 3.3

Let $S^{\in}$ have more than one image state, that is S has a number of consecutive states and all of them are on a non-optimal goal path to S_k. However, each such image state is at the same time on an optimal goal path to at least one other goal state.

As already described, the GSO proceeds as follows: (1) it chooses the image states that belong to the optimal goal path with the highest priority, and (2) amongst these image states (all on the same optimal goal path) it chooses the image state with the lowest state-index. Since the set of consecutive states of S is fixed the GSO will always choose one and the same image state from this set, say S'. Thus every time the GSO starts from S, it will always follow the state transition $S \rightarrow S'$. The image state S' is consequently materialized and the GSO proceeds from STEP 1.

If from STEP 1 it happens so that the GSO arrives at either STEP 2.2 or STEP 2.3 then we are guaranteed the achievement of G_k via one and the same non-optimal goal path $S \rightarrow S' \rightarrow ... \rightarrow G_k$.

If from Step 1 the GSO arrives again at Step 3 then we will have again either STEP 3.2 or STEP 3.3 which in turn guarantee the choice of one and the same consecutive image state S'' of S'. Thus the state transition sequence $S \rightarrow S' \rightarrow S''$ will be repeated every time the GSO starts from S.

Finally, since the GSO is following one and the same non-optimal goal path to G_k, this path will eventually cross the optimal goal path to this same goal state. In this case the GSO will choose (According to STEPS 2.2 and 2.3) a state on the optimal goal path to G_k and continue from this state towards G_k via the optimal goal path to G_k (according to Proposition 3:1).

Q.E.D.

3.5 The GSO In A Problematic Control Situation

In this section we identify a problematic control situation with a de-synchronization from the currently interpreted optimal goal path, where de-synchronization is defined as a transition from the currently interpreted goal path to another goal path. So far, a transition from one state to another was defined as the result of the occurrence of expected, or unexpected actions. In this section we introduce two additional causes for de-synchronization, namely, ill-represented states and violation of ontological assumptions. Then we describe the behavior of the GSO in the case of de-synchronization due to these different causes.

3.5.1 De-Synchronization

Let the present goal state be G_k and the object controller be on the optimal goal path to this goal state. At t-1 we have that $y_i(t$-$1)=1$ and the control action $u_{i,j}$ is executed, i.e., $u_{i,j}(t$-$1)=1$. Thus, at t-1 it is expected that the plant output y_j will be interpret as true at t.

At time t, the object controller reads certain sensory signals from the plant and based on these signals identifies the actual plant output as y, that is, $y(t)=1$. If the object controller works according to expectations then the expected plant output y_j and the actually observed one, y, must be equivalent, i.e., $(y \equiv y_j)$.

As already described, given that the expected plant formula y_j is interpreted as true at t, a number of control actions, $\{u_{j,k}\}$, exist which the object PC can potentially execute at t since they all have the same precondition, y_j. In this case the object PC uses the GSO to determine a control action which belongs to $\{u_{j,k}\}$, say $u_{j,r}$, and which when executed at t is expected to change the plant output from y_j to y_r, i.e., it is expected that $y_r(t+1)=1$.

However, it may be the case that the actual plant output at $t+1$, say y_k, is different from the expected one, y_r. That is, $y_r(t+1)=0$ and $y_k(t+1)=1$. Thus, the unexpected materialization of y_k implies a deviation from the optimal goal path to G. This deviation happens because the GSO guarantees that once the object controller is on the optimal goal to G_k and it will continue following it if at $t+1$ the plant formula interpreted as true at this sampling instant is part of some consecutive state of $(y_j, u_{j,r})$ (the controller state at t). In other words, the plant formula interpreted as true at $t+1$ must have the index r. However, the plant formula interpreted as true at $t+1$ is y_k and $r \neq k$. This implies that the state of which y_k is part of cannot be amongst the consecutive states of $(y_j, u_{j,r})$ and thus, cannot be on the optimal goal path to G. Hence, when y_k is completed with a control action, say $u_{k,n}$, we will have a transition from the image state $(y_j, u_{j,r})$ to the image state $(y_k, u_{k,n})$ instead of a state transition from $(y_j, u_{j,r})$ to some of its consecutive states. This particular situation is called *state de-synchronization* (or simply *de-synchronization*), and the expected plant output y_r is called a *failed plant output*. In short, a de-synchronization takes place any time when the plant formula materialized at t is not in the plant formula part of none of the states consecutive to a given state at t-1.

For example, consider the case where a controller commands a car-engine into a forced acceleration. Instead of the expected higher speed (the expected car-engine output), the engine may stop altogether if it cannot accommodate the level of acceleration commanded by the controller. Although this situation (stopped engine) may not have been foreseen as a possible car-engine output during acceleration, the controller should recognize the current car-engine output (stopped-engine) and begin controlling from there, say, commanding an engine-start action.

3.5.2 Typical De-Synchronization Causes

In what follows we will try to list some typical causes for de-synchronization which are often encountered in applications with PCs and indicate whether the GSO as described in Section 3.4.2 of this chapter can be used for the purpose of synchronization.

Let us note here that the de-synchronization can occur in both the continuous feedback and the sequential control parts of an object PC. While in the case of continuous feedback control, the de-synchronization is a well recognized phenomenon and there are well known ways of dealing with some cases of de-synchronization, this is not at all the case in sequential control. We are unaware of any systematic attempt to define de-synchronization and ways of dealing with it in the existing literature on hybrid control, discrete supervisory control, etc.

De-synchronization due to expected external actions

De-synchronization due to expected external actions is called *expected de-synchronization*. In this case the expected plant formula y_j does not materialize after the execution of a control action $u_{i,j}$ because of the occurrence of an expected external action $u_{i,k}^{ext}$. Such a de-synchronization may or may not occur depending on whether an expected external action occurs or not.

Let us have that the image state at t is $(y_i, u_{i,j})$, the goal state pursued is G_k, and $(y_i, u_{i,j})$ is on the optimal goal path to G_k. Let the control action $u_{i,j}$ be executed at t, given that $y_i(t)=1$. Thus, the expected effect of $u_{i,j}$ at $t+1$ is y_j. However, an external expected action $u_{i,k}^{ext}$ occurs at t. At $t+1$ it is then observed that the expected post-condition y_k of the $u_{i,k}^{ext}$ has overridden y_j, i.e., $y_k(t+1)=1$ and $y_j(t+1)=0$. Thus, the partial state $(y_k, -)$ is obtained at $t+1$. Consequently, y_k will not belong to the plant formula part of none of the states in the set of consecutive states image states of (y_i, $u_{i,j}$).

From the above it is easily seen that a de-synchronization can be easily identified as caused by an expected external action. The occurrence of such an action at t can be discovered at the next sampling instant because its expected effect upon the plant is specified a priori with a particular plant formula which materializes at $t+1$.

The GSO as presented in Section 5.2 of this chapter can be used for synchronization after an expected de-synchronization.

De-synchronization due to unexpected external actions

De-synchronization due to unexpected external actions is called *unexpected de-synchronization*. In this case the expected plant formula y_j does not materialize after the execution of a control action $u_{i,j}$ because of the occurrence of a unexpected external action. Such a de-synchronization may or may not occur depending on whether a unexpected external action occurs or not.

Suppose that at t the image state (y_i, $u_{i,j}$) is materialized on the optimal goal path to the present goal state G_k. Thus, when $u_{i,j}$ is executed at t it is expected that $y_j(t+1)=1$. However, suppose that at $t+1$ we have that $y_k(t+1)=1$ and $y_j(t+1)=0$, where $k \neq j$. Furthermore, let there be no expected external action $u_{i,k}^{ext}$ which if have been occurred at t, given that $y_i(t)=1$, would have been resulted in $y_k(t+1)=1$. Consequently, y_k will not belong to the plant formula part of none of the states in the set of consecutive states image states of (y_i, $u_{i,j}$).

As already described in Section 3.1.6 of this chapter, this situation indicates an occurrence of an unexpected external action. The GSO as presented in Section 5.2 of this chapter can be used for synchronization after an expected de-synchronization.

A de-synchronization due to the occurrence of an unexpected action can be distinguished from a de-synchronization due to the occurrence of an expected external action. This is possible since the latter one results in the materialization of an a priori specified plant formula, while the former can result in the materialization of any plant formula.

In continuous feedback control, if the *internal model principle* is used, the object controller is able to extend the plant model in a certain prescribed way so that it can determine an extended set of image states where some of these image states contain y_k [2].

De-synchronization due to ill-represented plant formula

As already described, the expected post-condition y_j of the control action $u_{i,j}$ executed at t may interpret as true at $t+1$ if y_i ($u_{i,j}$'s precondition) is already interpreted as true at t, i.e., $y_i(t) = 1$. If $u_{i,j}$ is executed given that $y_i(t) = 0$ then the expected post-condition y_j will not interpret as true at $t+1$.

Suppose that y_i is interpreted to true at t using some outdated plant signals, rather the actual plant signals at this sampling instant. This implies that if y_i was to be interpreted with the actual plant signals at t it would have been interpreted to false. Hence, as explained in the above, y_j will not interpret as true at $t+1$ when $u_{i,j}$ is executed, because the precondition of $u_{i,j}$ is actually false at t. At the same time, since $u_{i,j}$ is executed it will act on the actual plant output whatever it may be and thus, will transform it to a different one. The transformed plant output can be identified with a plant formula at $t+1$. Since, $y_j(t+1) = 0$ while some other plant formula different from it is interpreted to true, we again have a de-synchronization.

When the above described situation is present, y_i is called an *ill-represented formula*, since it does not represent the actual plant output. Because the object PC is a dynamic system, it is necessary to make sure that the plant formula interpreted to true in the object PC corresponds to the actual plant output since if this is not the case, there will be a switching transient (in the case of continuous feedback control).

Different types of de-synchronization caused by ill-represented plant formulas are discussed in a later section from this chapter for the case of continuous feedback control. Similar results for sequential control are not available in the existing literature.

It is easily seen that a de-synchronization due to an ill-represented state cannot be distinguished from a de-synchronization due to an unexpected external action: both prevent the materialization of the expected plant formula, and both can materialize any other plant formula. However, a de-synchronization due to either one of this two causes can be distinguished from de-synchronization due to an expected external action since the latter one results in the materialization of an a priori specified plant formula.

De-synchronization due to VOA

De-synchronization due to VOA is called *ontological de-synchronization.* Suppose that at t the image state $(y_i, u_{i,j})$ is materialized on the optimal goal path to the present goal state G_k. Thus, when $u_{i,j}$ is executed at t it is expected that $y_j(t+1)=1$. However, suppose that at $t+1$ we have that $y_k(t+1)=1$ and $y_j(t+1)=0$. Furthermore, let there be no expected external action $u_{i,k}^{ext}$ which when executed at t, given that $y_i(t) = 1$, results in $y_k(t+1)=1$. Consequently, y_k will not belong to the set of consecutive image states of $(y_i, u_{i,j})$. Such a de-synchronization occurs every time when there is a violation of ontological assumptions.

It is easily seen that a de-synchronization due to VOA cannot be distinguished from a de-synchronization due to either a unexpected external action or ill-represented state: all prevent the materialization of the expected plant formula, and all can materialize any other plant formula. However, a de-synchronization due to either one of these three causes can be distinguished from a de-synchronization due to an expected external action since the latter one results in the materialization of an a priori specified plant formula.

Thus, making a distinction between these three causes for a de-synchronization is of extreme importance. However, we have seen so far that one cannot distinguish between VOA, unexpected external actions, and ill-represented states provided only with a material image state at t and a partial image state at $t+1$. This is why, in Section 3.6 of this chapter we will try to make this distinction by studying the overall behavior of the GSO rather than just a single transition between states.

3.5.3 De-Synchronization In Continuous Feedback Control

Some of the typical de-synchronization causes described in the previous section are often encountered in continuous feedback control and there are well-known ways of dealing with them. In what follows we will describe some of the most often encountered de-synchronization causes in continuous feedback control and the different ways synchronization (though this operation is not explicitly named as such) is done for each particular cause.

Bumpless transfer

A (feedback) controller, when started, is expected to synchronize to a plant independently of what is the current state of the plant, i.e., the object controller should not have built-in requirements as to what the initial state of the plant should be. For example, when the control mode is switched from manual to automatic the state of the controller must have the correct value because the controller is a dynamic system. If this is not the case then there will be a switching transient. A smooth transition is called *bumpless transfer* and there are well known ways to achieve it. In the case of analog controllers it is customary to handle bumpless transfer by introducing a tracking mode which adjusts the controller state so that it is compatible with the given inputs and outputs of the controller. For digital controllers with state-feedback, bumpless transfer is achieved by building an explicit observer into the control algorithm.

Reset windup

A reset windup or integrator saturation for a controller with integral action can occur if the controller output saturates (e.g., due to limitations on an actuator using the controller output) and the controller continues to integrate the error. Thus the output of the controller can then assume very large values, and it can take a long time to get it back to a normal value again. This problem is avoided automatically when the velocity form of the control algorithm is used since the integration stops automatically when the output is limited.

Initialization

A controller can miss certain plant outputs during certain time interval (e.g., due to damaged sensors, sudden stop, etc.). Thus, because the regulator is a dynamic system, it is important to set the controller state appropriately (initialize it) so that it synchronizes to the plant state read after the controller has again started sensing the plant output. If this is not done, there may be large switching transients as it was the case with bumpless transfer.

In conventional process control with PI-controllers the controller has one state only - namely, the integrator output. It is customary to initialize such a controller by operating it in manual mode until the plant output comes close to its desired value.

For control algorithms with an explicit observer the controller state may be initialized by keeping the control signal fixed for the time required for the observer to settle.

On-line controller parameter changes

When some controller parameters are changed (e.g., the integration time) there is a need to change the state of the controller as well since this state depends on the controller parameters. For example, a change of the integration time will cause a step in the control signal unless the integral part is zero. One way to obtain smooth performance in the case of parameter changes is to store a set of past input/output data and to run an observer when the parameters are changed.

Time delays

A time delay makes information about the actual plant state to arrive later than desired to the controller. In this case the controller state at t will correspond to an old plant state observed at some previous to t time, rather than to the actual system state at time t.

Suppose a controller whose control action is to increase/decrease the concentration of a certain chemical. The concentration of the chemical is measured and the measurement procedure takes a time $T = 3$. The concentration is found too low at time 0 and the controller increases the dosage to increase the concentration. Any change due to a control action at time 0 will not be seen until time 3. Since the controller has not recorded any correction at time 1 it increases the dosage further, and continues to do so at time 2. The result is first observed at time 3. If the controller gain is too large, the concentration increase may have been too large. Consequently, the controller will increase the dosage, but will not see the result of this change until time 6, so it may further deteriorate the control at times 4 and 5.

The problem of controlling plants with delays can be solved with the use of the so-called Smith predictor which requires a model of both the plant and the delay.

Disturbances

Unexpected and expected external actions in continuous feedback control are known as disturbances. Disturbances, in the plant output, y, lead to control actions being taken based on a disturbed plant output rather than the undisturbed one. Disturbances on the controller output produce changes in the plant which do not correspond to the expected effect of the undisturbed controller output on the plant. It is customary to distinguish between three types of disturbances:

- Load disturbances: This type of disturbances influence the plant variables. For example, these may be disturbance forces in a mechanical system such as load on a motor, waves on a ship, etc., or in process control these may be quality variations in a feed flow or variations in demanded flow.

- Measurement errors: Measurement errors enter via the sensors. For example, there may be a steady state error in some sensors due to calibration, dynamic errors due to sensor dynamics, etc. In some cases it is not possible to measure the variable under control directly; the value of such a variable is inferred from indirect measurements of several other variables and the relationship between the controlled and measured variables can be quite complex (e.g., a non-linear time-varying relationship).

- Parameter variations: The disturbances appear as variations (inaccuracies) in the parameters of a linear model.

A change in the reference value (set point) is another type of disturbance. If a controller can use the advantage of information on the reference value change or any one of the above three types of disturbances can be modeled/measured as being of a particular type (e.g., impulse, pulse, step, ramp, and sinusoid), then this will generally improve the performance of the closed loop system. Feed forward control deals with this type of known disturbance (expected external actions). However, the quality of feed forward control depends heavily on the accuracy of both the disturbance models and measurements and of the process model. This means that any realistic implementation combines feed forward with feedback control. The feedback can compensate for inaccuracies in the process model, measurement errors, and unknown (unmodelled) disturbances since a corrective action occurs as soon the controlled variable deviates from the set point regardless of the source and type of disturbance. However, it cannot in a predictive way compensate for known disturbances.

The above described de-synchronization cases can be dealt with in relatively simple control applications, namely when the physical model of plant is well known, the discrete and continuous aspects of the plant can be separated, the control involves single-loops and a rather small number of input/output variables.

Control applications in hierarchical and distributed process control systems normally do not have the above mentioned characteristics: the plant is of a hybrid discrete/continuous nature, the control is plant-wide rather than focused on single loops, physical models are available for only certain plant phenomena, and there is a large number of input/output variables.

For such control applications, the overall control system consists of a large number of hierarchically organized PC units which during the operation of the overall control system may be stopped, become damaged, temporarily replaced by manual control from an operator, etc. At the same time the rest of the overall system is expected to continue the control in a way which is "consistent" with the current control situation for a particular PC. However, what is considered as consistent is very much application dependent and thus, the automated design of general-purpose goal seeking operations which can be directly used in an arbitrary control application and are robust with respect to de-synchronizations is an extremely difficult problem. The following example shows the difficulties encountered in a de-synchronization due to initialization.

Let us consider again the electric generators example. Now suppose that, say gen-5 is restarted after maintenance and its PC reads the current image states characterizing the PCs for the rest of the generators:

- gen-1 has failed to start for a second time

- gen-2 is delivering

- gen-3 is unloading

- gen-4 is synchronizing its frequency to the busbar

If the present electric load demand is high, what shall the goal seeking operation of the controller of gen-5 do? Since only gen-2 is delivering, gen-5 should start. However, before gen-5 begins delivering, either gen-1 or gen-4 may start delivering and then gen-5 should stop. However, since gen-3 is currently unloading, this means that to only start one generator is not enough to fulfill the load demand. Therefore if gen-1 fails for a third time, which is very likely, gen-5 has to start anyhow, etc.

We don't intend to provide here an optimal solution for this problem. The purpose of the example is to show that the goal seeking operation can initialize properly only if it has state information about the whole system of generators, (i.e. the power set of all generator states) on which it has to reason while during normal operation, (i.e., after initialized correctly) it needs only the state set corresponding to the operation of one generator. Therefore initialization in the context of the above system is a much more complex task then the initialization of a single PID-controller.

3.6 De-synchronizations And The GSO

In this section we describe the behavior of the GSO in the following three cases of de-synchronization:

- expected de-synchronization.

- unexpected de-synchronization.

- ontological de-synchronization.

- ill-represented state de-synchronization.

Furthermore, we generalize the behavior of the GSO in terms of different types of transition sequences where each type of a transition sequence is due to combinations of the above types of de-synchronization. We conclude by describing the ways for distinguishing between the different types of GSO behavior and the causes for it.

3.6.1 The GSO And Expected De-Synchronization

Let the present goal state be G_k and let at t the image state $S = (y_i, u_{i,j})$ on the optimal goal path to G_k be materialized. Thus, $u_{i,j}$ is executed at t, given that $y_i(t) = 1$, and it is expected that y_j will be interpreted as true at $t+1$.

Let there also be an external action $u_{i,k}^{ext}$ which occurs at t. That is, at $t+1$ we have that $y_k(t+1)=1$ and $y_j(t+1)=0$, and the GSO has to complete the partial state $(y_k, -)$ instead of the expected one $(y_j, -)$. As already explained in the previous section this is the case of expected de-synchronization.

Since y_k is a plant formula which is not in the plant formula part of none of the consecutive states of $(y_i, u_{i,j})$, $(y_k, -)$ cannot be completed by using one of these image. Obviously, to complete this partial state the GSO should consider the set of consecutive states of the well-defined state $S' = (y_i, u_{i,k}^{ext})$. This state is obviously not an image state, i.e., it does not belong to an optimal goal path, for the simple reason that $u_{i,k}^{ext}$ is an external action and not a control action. Thus, states like $S' = (y_i, u_{i,k}^{ext})$ are "bridges" between optimal goal paths or non consecutive states on the same optimal goal path: y_i is in the plant formula part of an image state on the optimal goal path to G_k and y_k (the post-condition of $u_{i,k}^{ext}$) is in the plant formula part of another image state on the same optimal goal path or a different one. The fact that y_k is in the plant formula part of another image state on the same optimal goal path or a different one follows from our description of external actions in Section 3.1.6 .

Consequently, every time such a state is materialized while following the optimal goal path to G_k, it will not belong to the set of consecutive states of the previously materialized image state. However, its set of consecutive states cannot be empty (unless it is a goal state) and will contain image states from the optimal goal path to G_k or some other optimal goal path to a different goal state.

With respect to the set of consecutive image states of $S' = (y_i, u_{i,k}^{ext})$ the following cases can be distinguished.

CASE 1

The set of consecutive states of $S' = (y_i, u_{i,k}^{ext})$ contains at least one image state on the optimal goal path to G_k;

CASE 2

The set of consecutive states of $S' = (y_i, u_{i,k}^{ext})$ does not contain image states on the optimal goal path to G_k. In this case two further mutually exclusive subcases can occur.

CASE 2.1

The set of consecutive states of $S' = (y_i, u_{i,k}^{ext})$ contains at least one image state on a non-optimal goal path to G_k;

CASE 2.2

The set of consecutive states of $S' = (y_i, u_{i,k}^{ext})$ contains no image state on a non-optimal goal path to G_k;

In any of the above cases the GSO, when trying to complete the partial state $(y_k, -)$ obtained after the execution of $u_{i,k}^{ext}$, will start by first constructing the set of consecutive states of $S' = (y_i, u_{i,k}^{ext})$ which will contain image states not an the optimal goal path to the current goal state G_k. Thus, the GSO will behave as stated in Proposition 3:2 : CASE 1 corresponds to STEP 2.2 or STEP 2.3; CASE 2.1 corresponds to STEP 2.1 followed by STEP 3.2; Case 2.2 corresponds to STEP 2.1 followed by STEP 3.3. This in turn implies, that after an occurrence of expected de-synchronization, i.e., a de-synchronization due to an expected external action, the continued operation of the GSO will result in one of the following scenarios:

- G_k is achieved by following the optimal goal path to it;

- G_k is achieved by following a non-optimal goal path to it;

- G_k is not achieved, but some other goal state, say G_l, whose priority is lower than this of G_k is achieved.

- G_k is not achieved because of cyclic behavior (expected cycle) of the GSO.

The occurrence of additional expected de-synchronizations due to other expected external actions may occur while the interpretation operation is materializing the image states obtained at the different steps of the operation of the GSO, and then again one of the CASES *1*, *2*, or *3* can occur eventually leading to one of the above described scenarios.

However, in the last scenario from above, the GSO starts repeating certain state transition sequence (expected cycle) when the following situation occurs. We will illustrate this cyclic behavior for CASE 1, but it may be encountered in any one of the other cases as well. In CASE 1, we will have that the set of consecutive image states of $S' = (y_i, u_{i,k}^{ext})$ contains at least one image state on the optimal goal path to G_k. Suppose that this image state, say $S''=(y_k, u_{k,q})$ has the lowest state-index amongst all image states in S' and also, a lower state index than $S=(y_i, u_{i,j})$.

Recall here that $S=(y_i, u_{i,j})$ is on the optimal goal path to G_k, it was materialized at t, and again at t $u_{i,k}^{ext}$ occurred. Thus at $t+1$, y_k was interpreted as true instead of y_j.

The GSO, will choose the state with the lowest state index in S', i.e., $S'' = (y_k, u_{k,q})$. Since this state is on the optimal goal path to G_k, the GSO will continue towards G_k on this optimal goal path as long as no further expected de-synchronizations occur according to Proposition 3:1. If the latter is the case the GSO will eventually materialize again the image state $S=(y_i, u_{i,j})$ since it has a greater state-index than $S'' = (y_k, u_{k,q})$. If now $u_{i,k}^{ext}$ again occurs then at the next sampling instant it will be again the partial state $(y_k, -)$ that has to be completed instead of $(y_j, -)$. Thus, the state transition sequence $S' \rightarrow ... \rightarrow S'' ... \rightarrow S$ will be repeated every time $u_{i,k}^{ext}$ is executed together with $u_{i,j}$. Since the GSO does not have any information about cycles that include the repeated occurrence of a control action and an external action, it will not be able to leave this cycle and thus never reach G_k.

Let us note here that the occurrence of $u_{i,k}^{ext}$ every time the image state $S = (y_i, u_{i,j})$ is materialized may be highly unlikely. However, since the behavior of the object PC can hardly be left to chance there is, a need to distinguish this type of cycles from normal cycles and to be able to leave the former cycles in an appropriate manner. The case when $u_{i,j}$ may be affected by (or may affect) somehow u_{ik}^{ext} is discussed in Chapter 4.

3.6.2 The GSO And Unexpected De-Synchronization

Let the present goal state be G_k and let at time t the image state $S = (y_i, u_{i,j})$ on the optimal goal path to G_k be materialized. Thus, $u_{i,j}$ is executed at t, given that $y_i(t)=1$, and it is expected that y_j will be interpreted as true at $t+1$.

Let there also be no external action $u_{i,k}^{ext}$ which can be executed at t. However, let at $t+1$ we have that $y_j(t+1) = 0$ and $y_l(t+1) = 1$. As already described in Section 3.1.6 this indicates the execution of an unexpected external action and since y_l cannot be in the set of consecutive states of $S = (y_i, u_{i,j})$ we have a de-synchronization caused by an unexpected external action. Thus at $t+1$ the GSO has to complete the partial state $(y_l, -)$, instead of $(y_j, -)$.

In this case we have that a state like (y_i, u^{unexp}) has been materialized at t. This however, is not a well-defined state since we do not have symbols for unexpected actions. Despite this, since y_l is a plant formula encountered in the plant formula part of at least one other image state, the set of consecutive states of (y_i, u^{unexp}) cannot be empty. Because of this, we again have one of the CASES 1, or 2 from the previous subsection, and the GSO will proceed in exactly the same manner as in the case of de-synchronization caused by an expected external action. Consequently, the continued application of the GSO will again result in one of the following scenarios:

- G_k is achieved following the optimal goal path to it;

- G_k is achieved following a non-optimal goal path to it;

- G_k is not be achieved, but some other goal state, say G_l, whose priority is lower than this of G_k is achieved.

- G_k is not achieved because of cyclic behavior (unexpected cycle) of the GSO.

In the last one of the above scenarios the GSO may again start repeating certain state transition sequence (cyclic behavior) when (CASE 1) the completed version $(y_l, u_{l,n})$ of the partial state $(y_l, -)$ is on the optimal goal path to G_k, has a lower state-index than $S = (y_i, u_{i,j})$ and the same unexpected action occurs simultaneously with $u_{i,j}$. Since the GSO does not have any information about cycles that include the repeated occurrence of a control action and unexpected external action, it will not be able to leave this cycle and thus never reach G_k.

The occurrence of a cycle of the above type is even less unlikely than the occurrence of a cycle caused by expected external actions, but as long as there is a remote possibility for this, there is a need for the GSO to distinguish this type of cycle from normal cycles and expected external actions cycles, and to be able to leave them in an appropriate manner.

3.6.3 The GSO And Ontological De-Synchronization

The case of ontological de-synchronization occurs when y_i is interpreted as true at t, but because of the violation of some ontological assumptions, it does not represent any physically possible plant output. Thus, the control action $u_{i,j}$ when executed, will not realize its expected effect upon the plant. To illustrate this situation let us consider a number of examples. In the first example, a violation of ontological assumptions occurs in the physical components of the plant. In the second example, a violation of ontological assumptions occurs in a continuous feedback control context, and in the third example, it occurs in the context of sequential control. In all of these examples the object PC is referred to as the controller.

Example 1

This example uses a variation of the example with the diesel engine - generator in **Figure 8**. The plant consists now of a diesel engine and a generator connected by a clutch. The object PC uses a proximity switch on a tooth wheel to determine the speed of the engine for low speed levels and the frequency of the generator to measure the rotation speed at normal speed levels.

A typical start-up sequence of the diesel engine may be repeated since the engine may not start at the first try. A start sequence of the diesel engine - generator unit consists of the sequences of state transitions shown in **Figure 14**.

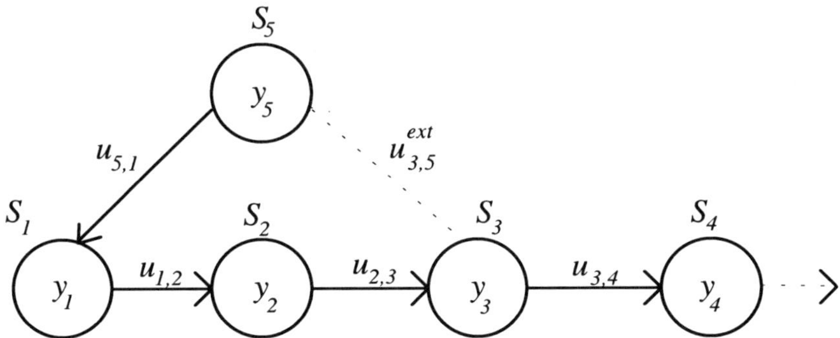

Figure 14 Start sequence for a Diesel generator.

The states are the following from:

S_1 - the controller waits a time interval T_1 before it performs a next (or the first) start of the diesel engine.

S_2 - the controller executes a new engine start

S_3 - the controller waits for a time T_2 required for the engine to gain speed

S_5 - the diesel engine has failed to start

S_4 - the engine has started (goal state)

The plant formulas use the following constants:

N_{lim} - is the speed limit (in rpm) the controller uses to change the speed measurement. If the engine speed is under N_{lim} then the speed takes the value N_1 from the proximity sensor, else the value N_2 computed from the frequency of the generator.

T_1 - time interval required until a new engine start operations can be triggered

T_2 - time interval required for the engine to accelerate from 0 to N_{lim} during start-up.

M_{max} - the maximum number of allowed start-up operations

The following dynamic variables are used:

F_{count} - current number of engine start-up repetitions

t_1 - timer for the interval T_1

t_2 - timer for the interval T_2

The plant formulas and the control actions for the states $S_1,...,S_5$ are the following:

$$S_1 = (y_1, u_{1,2})$$

$$y_1 = (N_1 < N_{lim}) \wedge (F_{count} < M_{max}) \wedge (t_1 < T_1)$$

$$u_{1,2} = \text{`` } t_1 = t_1 + \Delta t \text{ ''}$$

(y_1 is interpreted as true when the engine speed is lower than N_{lim}, the number of start attempts is under the maximum limit and the time to start a new attempt is not yet reached. The control action is to update the timer t_1).

$S_2 = (y_2, u_{2,3})$

$y_2 = (N_1 < N_{lim}) \wedge (F_{count} < M_{max}) \wedge (t_1 \geq T_1)$

$u_{2,3} = $ *"start_engine" and "reset t_1" and "reset t_2"*

(y_2 is interpreted as true when the engine speed is lower than N_{lim}, the number of start attempts is under the maximum limit but the time limit for new attempt to start the engine is reached. The corresponding control action is to trigger a start command for the engine and to reset the timers).

$S_3 = (y_3, u_{3,4})$

$y_3 = (N_1 < N_{lim}) \wedge (t_2 < T_2)$

$u_{3,4} = $ *" $t_2 = t_2 + \Delta t$ "*

(y_3 is interpreted as true when the speed is lower than N_{lim} and the time until the engine should gain speed is not over. The timer t_2 is updated and the controller is waiting for the outcome of the start-up operation)

$S_4 = (y_4, u_{4,6})$

$y_4 = (N_2 \geq N_{lim})$

$u_{4,6} = $ *"start engine speed regulator"*

(y_4 is interpreted as true when the engine has gained a speed larger than N_{lim} which is measured from the frequency of the generated electrical current, N_2. The action is to start a speed regulator.)

$S_5 = (y_5, u_{5,1})$

$y_5 = (N_1 < N_{lim}) \wedge (t_2 \geq T_2)$

$u_{5,1} = $ *'reset t_2'*

(y_5 is interpreted as true when the engine speed is under N_{lim} and the time interval during which the engine should have started, T_2, is over. In other words, the state S_5 is true when the engine has failed to start within the time interval N_{lim}. Note that the state S_5 materializes as a consequence of the occurrence of the external action $u_{3,5}^{ext}$ which is the unexpected action of the engine start failure.)

The intended state sequences for the start of the diesel engine are the following:

$S_1 \rightarrow S_2 \rightarrow S_3 \rightarrow S_4$ - the engine is started successfully

$S_1 \rightarrow S_2 \rightarrow S_3 \rightarrow S_5 \rightarrow S_1$ - the engine fails to start

One of the ontological assumption for this control is that *the rotation speed of main axle of the diesel engine is equal to the rotation speed of the main axle of the generator*. Let us assume now that this ontological assumption does not hold. For instance the diesel engine is started but the clutch between the axle of the engine and that of the generator is open due to a mechanical fault. The sequence of states is then the following:

- S_1 materializes, the controller waits

- S_2 materializes, the controller triggers the engine start command and assume that the engine starts

- S_3 materializes, the controller waits for the engine to speed-up

However, after the state S_3, the controller is de-synchronized since:

- The state S_4 cannot materialize since the speed N_2 is zero (the generator axle does not rotate and the frequency is zero).

- The state S_5 cannot materialize either since the speed N_1 is greater than N_{lim} (the engine has reached a high rotating speed).

Therefore the controller cannot materialize any of the consecutive states of the state S_3.

Example 2

The following example shows a radar tracking control system which has two positioning mechanisms. First, a rough position of the target object is given to the radar and a regulator R1 brings the radar into this position. For this purpose, the regulator measures the position using a mechanical device. After the rough position is reached, another regulator, R2, locks the radar onto the target using an optical tracking principle which does not require any more the mechanical positioning measurement. The ontological assumption is that *when the regulator R1 has reached the set-up position, then the object is in the locking range of the regulator R2.* However, a violation of this assumption occurs: the mechanical measuring device is not rigidly fixed to the radar axle and has an offset error. Therefore instead of the real angle α it gives an angle $\alpha+\varepsilon$. The state transitions are shown in the **Figure 15**.

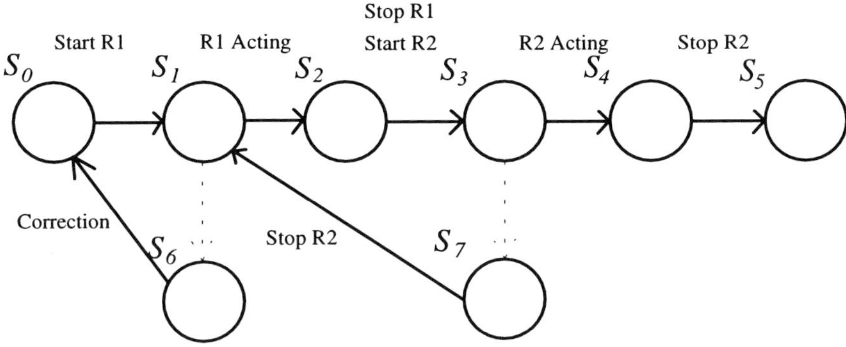

Figure 15 State set of a radar tracking system (i).

The states are the following:

S_0 y_0 = (reference value received to start tracking)
$u_{0,1}$ = (start regulator R1)

S_1 y_1 = (in range for R1) $\wedge\neg$ (in range for R2)
$u_{1,2}$ = (regulator action R1)

S_2 y_2 = (in range for R2) $\wedge$ (R1 is active)
$u_{2,3}$ = (stop regulator R1) $\wedge$ (start regulator R2)

S_3 y_3 = (in range for R2) $\wedge \neg$ (range for goal)
$u_{3,4}$ = (regulator action R2)

S_4 y_4 = (in range for goal)
$u_{4,5}$ = (stop R2) $\wedge$ (start goal)

S_5 y_5 = (goal fulfilled)
$u_{5,5}$ = (wait)

S_6 y_6 = (outside position for R1)
$u_{6,0}$ = (some correction action)

S_7 $y_7 =$ (outside working position for R2)

$u_{7,1} =$ (stop R2)

The sequence of state transitions is the following:

- The system start at S_0 with a reference value for the regulator R1.

- At the state S_1 the regulator R1 brings the radar into the reference position.

- The state S_2 stops R1 and starts R2.

- The ontological assumption is that the state S_3 can always be materialized after S_2 since the radar is now in range and R2 can be started.

- This assumption is not true due to the measurement device offset, which gives a large position error, identical to one obtained after an external disturbance. Therefore the state S_3 cannot materialize. However, the state S_2 cannot materialize either since, the regulator R1 is not active any more. A state that can materialize is S_7 since the offset error, is read as a large control error. But S_7 is not consecutive to S_2 and no other state consecutive to S_2 can materialize.

Example 3

A train connection ties three cities A, B, C and is controlled remotely (**Figure 16**). The controller actions are to start and stop the train at the stations and to close and open a barrier placed at a passageway between the cities B and C. Moreover, the train has a local safety breaker which stops the train in B if the barrier is open. The controller senses the position and the speed of the train using six measurement points, $P_1,...,P_6$ placed on the railway.

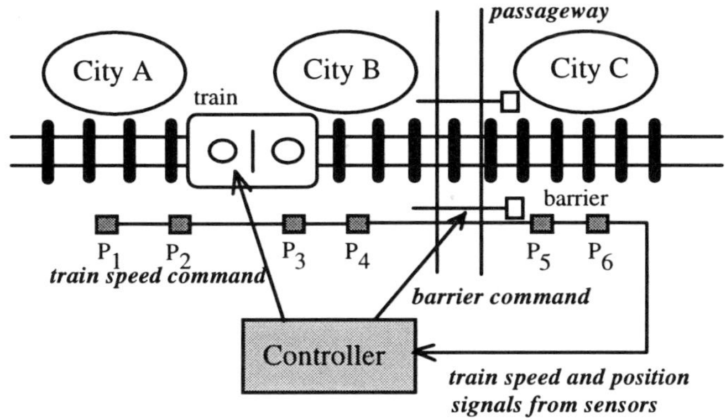

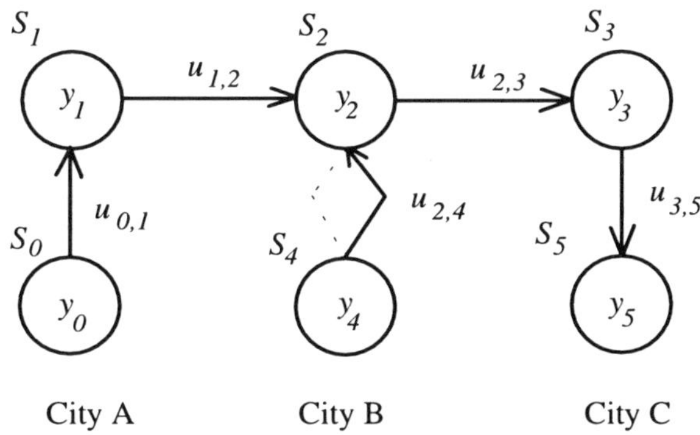

Figure 16 Train traffic controller.

The specified control sequence is the following. The train starts from A. When the train reaches the cruise speed outside A, the controller closes the barrier placed between B and C. When the train arrives near B, if the barrier is open the train stops in B, otherwise it proceeds towards C. When the train arrives near C, the controller begins to break the train and opens the barrier. When the train is in C, the controller stops it.

The state set of the controller is the following:

$S_0=(y_0, u_{0,1})$ - the train is stopped in A, speed and position received from P_1

$y_0=(\text{speed}=0) \wedge (\text{position}=A)$

$u_{0,1}=$ "start train"

$S_1=(y_1,\ u_{1,2})$ - train running with cruise speed out of A, speed and position received from P_2

$y_1=$ (speed=cruise) $\wedge$ (position=A)

$u_{1,2}=$ "control train speed" and "close barrier"

$S_2=(y_2,\ u_{2,3})$ - train arriving with cruise speed to B, speed and position received from P_3

$y_2=$(speed=cruise) $\wedge$ (position=B) $\wedge$ (barrier=closed)

$u_{2,3}=$"control train speed"

$S_3=(y_3,\ u_{3,5})$ - train arriving with cruise speed to C, speed and position received from P_5

$y_3=$(speed=cruise) $\wedge$ position=C)

$u_{3,5}=$'start to break' and 'open barrier'

$S_4=(y_4,\ u_{4,2})$ - train stopped in B, speed and position received from P_4

$y_4=$(speed=0) $\wedge$ (position=B)

$u_{4,2}=$"start train" and "close barrier"

$S_5=(y_5,\ \_)$ - train stopped in C, speed and position received from P_6

$y_5=$(speed=0) $\wedge$ (position=C)

The external action $u_{2,4}^{ext}$ is the local break of the train (independent of the controller) which stops the train in B if the barrier is open.

From the state description above, it can be easily seen that when the train travels from A to C, the controller follows the state sequence $S_0 \rightarrow S_1 \rightarrow S_2 \rightarrow S_3 \rightarrow S_5$.

Let us see now what is the behavior of the controller in cases of violations of ontological assumptions. Let us assume that there is an error in how the signals are read from the plant, such that the position of the train in P_3 is mixed up with the position of the train in P_5. That means, when the train is in B running with cruise speed, the controller reads that the train is in C and vice versa. However, when the train is stopped in B or C, then the train position is read right. The state transitions are then the following. When the train starts from A, the state transition $S_0 \rightarrow S_1$ materializes as specified. However, due to the position reading error at P_3, after the state S_1, the state S_2 does not materialize any more but the state S_3 materialized instead. Since the state S_3 is not consecutive to S_1, it means that in S_1 we have a de-synchronization due to a VOA on state S_2.

Note that if the controller continues to control from the new state S_3, the following happens. The controller is in the state S_3, however in reality the train is in B and not in C. That means the proper state would be S_2 and not S_3. Since the control action at S_3 is a break command to the train and an open command to the barrier, that in effect stops the train. Since the reading at P_4 is now right, the next state materialized after S_3 will be not the expected one S_5 but S_4. Therefore the control has a new de-synchronization. The controller in the state S_4 executes a train start and a barrier close command, but due to the signal error, when the train reaches the cruise speed, the controller will read the state S_3 instead of the expected S_2. Finally, due to the external action $u_{2,4}^{ext}$, after each attempt of the train to start from S_4, the barrier will be closed and then the train stopped and the barrier open. The controller will be locked in this loop.

Now let us return to the case of de-synchronization caused by violation of ontological assumptions.

After such a violation, the expected plant formula y_j does not materialize at $t+1$ since it no longer represents a physically feasible plant state. Instead of it, a different plant formula y_k is materialized which is not in the plant formula part of none of the consecutive states of the state materialized at t. Since this is exactly the situation due to the occurrence of an unexpected external action we will have again the following for the behavior of the GSO:

- G_k is achieved following the optimal goal path to it;

- G_k is achieved following a non-optimal goal path to it;

- G_k is not be achieved, but some other goal state, say G_l, whose priority is lower than the priority of G_k is achieved.

- G_k is not achieved because of cyclic behavior (ontological cycle) of the GSO.

In the last case from above the GSO is forced in a cyclic behavior which starts and ends with the image state whose plant formula is affected by the violation of ontological assumptions. This cycle will be repeated an indefinite number of times, if the violation persists indefinitely.

3.6.4 The GSO And Ill-Represented State De-Synchronization

After such a de-synchronization, the expected plant formula y_j does not materialize at $t+1$ since the materialized plant formula at t does not represent the actual plant state at this sampling instant. This was due to the fact that y_j was interpreted to true using outdated plant signals. Thus, y_j is actually false at t and the control action $u_{i,j}$ which has this plant formula as its precondition cannot achieve its expected effect upon the plant. Instead of the expected plant formula y_j, a different plant formula y_k is materialized which is not in the plant formula part of none of the consecutive states of the state materialized at t. Since this is exactly the situation due to the occurrence of an unexpected external action or a VOA we will have again the following for the behavior of the GSO:

- G_k is achieved following the optimal goal path to it;

- G_k is achieved following a non-optimal goal path to it;

- G_k is not be achieved, but some other goal state, say G_l, whose priority is lower than the priority of G_k is achieved.

- G_k is not achieved because of cyclic behavior (ill-represented cycle) of the GSO.

From the results presented in this section it is obvious that despite of the different nature of the causes for a de-synchronization the overall behavior of the GSO in the case of each particular cause is the same. However, the case when a goal state cannot be achieved due to the cyclic behavior of the GSO is of a particular interest. First, because once the GSO is in a cycle not only a particular goal state cannot be achieved, but none of the goal states can be achieved. Second, one may study the type of transitions involved in a cycle and on the basis of this identify different types of cycles, e.g., whether the cycle consists only of state transitions, or of state transitions and transitions due to expected external actions, etc. Then being able to distinguish between different types of cycles one might be able to distinguish between the different causes for de-synchronization

3.6.5 Generalizing Cyclic Behavior

Let $S = (y_i, u_{i,j})$ be the image state materialized at t and on the optimal goal path to G_k. Thus, we have that $u_{i,j}$ is executed at t, given that $y_i(t) = 1$ and y_j is expected to interpret as true at $t+1$. However, at $t+1$ we obtain $y_j(t+1)=0$ and $y_k(t+1)=1$. Thus the GSO has to complete the partial state $(y_k, -)$.

Suppose that the above partial state is completed as $S' = (y_k, u_{k,l})$ and there are no more de-synchronizations. The condition which guarantees the cyclic behavior of the GSO, i.e., the GSO will eventually arrive again at S, is as follows: at least one image state on the optimal goal path to G_k is reachable from S' and this image state has a state index lower than this of S .

If this condition holds it is easy to verify, by using Propositions 3:1 and 3:2 and CASES 1, 2, and 3 from Section 3.6.1, that the GSO will return to this image state, say S''. Thus, the state transition sequence $S \rightarrow S' \rightarrow {}_{...} \rightarrow S'' \rightarrow S$ will be repeated as many times as the transition $S \rightarrow S'$ materializes.

If there are further de-synchronizations occurring after the materialization of S' then in order to return back to S (i.e., complete the cycle) the above condition has to hold for every image state obtained after de-synchronization.

Now let us identify types of state transitions and transitions from which a cycle can be built. Consider a transition $S \rightarrow S'$ where $S = (y_i, u_{i,j})$ and $S' = (y_k, u_{k,p})$ and let this transition be materialized during two successive sampling instants. We can distinguish between the following cases:

- When $j = k$ (i.e., $y_j \equiv y_k$ and $u_{k,p} \equiv u_{j,p}$) we have that $S \rightarrow S'$ is the result of a control action only and it is called a *control transition* that, is a control transition is a state transition.

- When $j \neq k$ (i.e., $y_j \neq y_k$ and $u_{k,p} \neq u_{j,p}$) and there exist an expected external action $u_{i,k}^{ext}$ the transition $S \rightarrow S'$ is the result of this control action and is called an *expected transition*, that is, an expected transition is a transition due to expected external action.

- When $j \neq k$ (i.e., $y_j \neq y_k$ and $u_{k,p} \neq u_{j,p}$) and there does not exist an expected external action $u_{i,k}^{ext}$ the transition $S \rightarrow S'$ is the result of either an unexpected external action, a violation of ontological assumptions, or ill-represented state and is called an *unexpected transition* that is, an unexpected transition is a transition due to either an unexpected action, a VOA, or an ill-represented state.

As already explained, these three different causes for an unexpected transition cannot be distinguished from each other.

Using the above types of state transitions let us now define the following types of cycles and sequences of transitions.:

- Control cycle: a cycle consisting only of control transitions; Control sequence: consists only of control transitions (e.g., goal paths). Such a cycle is also called a *normal control cycle* and it is defined a priori and is executed only a desired number of times.

- Expected cycle: a cycle consisting only of at least one expected transition and at least one control transition; Expected sequence: consists only of at least one expected transition and at least one control transition.

- Unexpected cycle: a cycle consisting only of at least one unexpected transition and at least one control transition. Unexpected sequence: consists only of at least one unexpected transition and at least one control transition.

- Mixed cycle 1: a cycle consisting only of at least one expected and at least one unexpected transition. Mixed sequence 1: consists only of at least one expected and at least one unexpected transition.

- Mixed cycle 2: a cycle consisting only of at least one expected, at least one unexpected, and at least one control transition. Mixed sequence: a sequence consisting only of at least one expected, at least one unexpected, and at least one control transition.

For each of the above types of cycles the following general characteristics can be used to distinguish between some of them:

Control cycles

As already mentioned in Section 3.4.4, the occurrence of such cycles is determined in advance and they are always repeated a desired (finite) number of times. Furthermore, each such cycle is identified with unique state transition sequence consisting only of control transitions. The GSO can enter, execute the desired number of times, and exit a control cycle. Thus, a control cycle can be distinguished from the rest of the types of cycles which do not consist only of control transitions.

Expected cycles

The occurrence of such cycles cannot be predicted in advance: such a cycle may or may not occur, and the probability of repeated occurrence is very low. The number of times an expected cycle can occur cannot be determined in advance either. Since such a cycle consists only of control and expected transitions which both are specified a priori it can be distinguished from a control cycle (only control transitions) and from the rest of the types of cycles which involve unexpected transitions .

Unexpected cycles

The occurrence of such cycles cannot be predicted in advance: such a cycle may or may not occur, and the probability of its repeated occurrence is extremely low. The number of times a unexpected cycle can occur cannot be determined in advance either. Since such cycles consist only of control and unexpected transitions and do not have expected transitions they can be distinguished from control cycles (only control transitions), expected cycles (at least one expected transition), mixed cycles *1* (at least one expected transition), and mixed cycles *2* (at least one expected and at least one control transition).

Mixed cycles 1

The occurrence of such cycles cannot be predicted in advance: such a cycle may or may not occur, and the probability of its repeated occurrence is extremely low. The number of times a mixed cycle *1* can occur cannot be determined in advance either. Since such cycles consist only of expected and unexpected transitions and do not involve control transitions they can be distinguished from control cycles (only control transitions), expected cycles (no unexpected transitions and at least one control transition), and mixed cycles *2* (at least one control transition).

Mixed cycles 2

The occurrence of such cycles cannot be predicted in advance: such a cycle may or may not occur, and the probability of its repeated occurrence is extremely low. The number of times a mixed cycle *2* cycle can occur cannot be determined in advance either. Since such cycles consist of expected, unexpected and control transitions they can be distinguished from control cycles (only control transitions), expected cycles (only control and expected transitions), and mixed cycles *1* (only expected and unexpected transitions).

However, one cannot distinguish between the causes for unexpected cycles, mixed cycles *1* and *2* since these involve unexpected transitions and as we already saw, the causes for a unexpected transition (unexpected actions, ill-represented states, and VOA) cannot be distinguished from each other.

Now if we turn to the first three scenarios describing the overall behavior of the GSO where despite of a de-synchronization the original or some other goal state is achieved, we can draw the following conclusion:

In all three scenarios, a goal state is achieved despite of a de-synchronization by either following an optimal goal path or a non-optimal goal path This implies that the goal state has been achieved by following either an expected sequence, unexpected sequence, or some of the mixed types of sequences. These can be distinguished from each other in the same manner as the corresponding types of cycles. However, the causes for a de-synchronization in the case of unexpected and mixed sequences still cannot be distinguished from each other because these sequences involve unexpected transitions and the causes for a unexpected transition cannot be distinguished from each other.

3.7 Summary

In this chapter we introduced the formal concepts for object PC-based control, such as states, state transitions, and transitions, optimal and non-optimal goal paths. Furthermore we described the control operations of an object PC, namely the operations of interpretation and state completion. The state completion operation is realized, in the general case, via the use of the goal seeking and synchronization operations. We provided a formal description of these two operations using the notion of an extended state. Then we showed two major properties of the goal seeking operation which in the absence of a de-synchronization lead always to the achievement of all of the goal states of the object PC. Furthermore, we identified a problematic control situation with a de-synchronization. A de-synchronization was described as a transition from the currently interpreted goal path to the same or another goal path, transition which is not among the state transitions the PC performs with its control actions. Then we described a number of causes for a de-synchronization, namely expected external actions, unexpected external actions, ill-represented states, VOA, and described the behavior of the GSO in different cases of de-synchronization (each de-synchronization case due to a particular cause). Furthermore, we classified the behavior of the GSO in terms of different types of transition sequences (cyclic and acyclic transition sequences) each type of a transition sequence involving transitions due to combinations of the different causes for a de-synchronization. Finally, we arrived at the conclusion that one can distinguish between the different types of behavior of the GSO in terms of the different types of transition sequence corresponding to each type of behavior. However, we also arrived at the conclusion that for certain types of behavior one cannot distinguish between the causes for it. Namely, whenever a behavior expressed in terms of unexpected and mixed transition sequences is observed the causes for it can be any one of unexpected actions, VOA, or ill-represented states, but one cannot specify exactly which one of these three causes is the actual one.

In the next chapter we will introduce an extended notion of a well-defined state, and will restrict the state set of the object PC so that one can distinguish between the above three causes for a de-synchronization.

REFERENCES

[1]   Ravn, A.P., Rischel, H., Hansen K.M. Specifying and Verifying Requirements of
       Real-Time Systems, IEEE Trans on Software Eng. January 1993

[2]   Wonham, W. M. Towards an Abstract Internal Model Principle. In IEEE Trans on
       Systems, Man and Cybernetics, vol SMC-6 no. 11, Nov 1976.

4

A WELL-DETERMINED STATE SET

In this chapter we extend the definition of a well-defined state with the concept of a *control configuration*. A control configuration allows for taking into account the actuator outputs the PC uses to act upon the plant. Thus, the effect of a control action is described in terms of a particular actuator output which in turn is expected to produce certain plant output. In this context, a plant formula is now a pair consisting of: (*i*) a configuration formula, describing an actuator output due to a control action, and (*ii*) a proper plant formula, describing the expected plant output due to an actuator output. Recall here, that a plant formula, as introduced in Chapter 3, does not reflect explicitly this distinction between actuator and plant outputs, but is in fact an aggregated representation of these two different types of outputs. A representation of a state which makes explicit the distinction between actuator and plant outputs allows us to formulate a number of constraints on the state set S. The so constrained state set imposes further constraints on the allowable transitions and state transitions which in turn eliminate the presence of non-optimal goal paths. Under these constraints, we show in Chapter 5 that a de-synchronization due to VOA can be distinguished from de-synchronizations due to unexpected external actions and ill-represented states. Thus, the goal of the present chapter is to introduce the control configuration related notions of *states*, *transitions* and *goal paths* which will be used in Chapters 5 and 6.

4.1 Control Configurations

In this section we introduce the concept of a control configuration by first discussing informally the actuator interface of an object PC. Then we describe formally a control configuration in terms of the so called configuration formulas and re-consider the notions of plant formulas, control and external actions in the context of the explicit representation of control configurations as configuration formulas.

4.1.1.   The Need For Control Configurations

An object PC acts upon the plant using physical devices called *actuators*. The actuators are the interface between the object PC and the plant it controls, according to the control scheme shown in **Figure 17** (the notation in the figure is defined in Section 3.3).

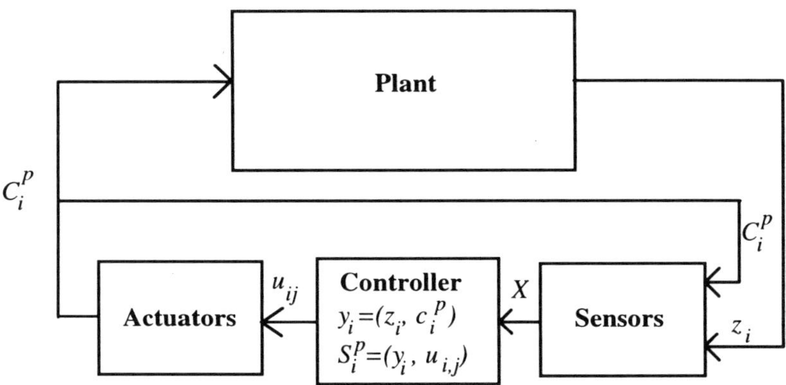

Figure 17 Control scheme with actuators.

An actuator has a number of physical outputs which are relevant for the control due to the effects they can have upon the plant. Each control action $u_{i,j}$ the object PC executes results in a specific actuator output. After the execution of a control action, the currently obtained actuator outputs persist until the object PC executes a new control action. An example of an actuator is an *electrical breaker*. The output of the electrical breaker is its electrical contact position: closed or open. Normally the object PC changes the breaker position using two binary outputs: one closes and the other one opens the breaker. After a *close* or an *open* control action, the breaker contact remains in its position until the PC executes a new control action. An example of an actuator with continuous output is a power amplifier together with the mechanical device it positions (i.e., a servomechanism). The relevant output of this actuator is the position of the target mechanical device. An actuator with continuous output and binary input is the valve positioning system shown in **Figure 18** . The input to the actuator (i.e., the control action) uses two binary command signals - one increases and the other one decreases the valve angle via a power amplifier. The latter drives an electrical motor and a mechanical device which transforms the rotation angle of the electrical motor into the rotation angle of the valve. The output of the actuator in **Figure 18** is the valve angle relative to a reference position.

An actuator may have more than one output. For instance, the actuator shown in **Figure 18** may be adjusted for the power factor of the power amplifier and for the mechanical conversion factor of the motor axle relative to the rotation of the valve. In this case, the outputs are the power factor, the mechanical conversion factor and the valve angle. The set of outputs of all the actuators of an object PC at the same sampling instant t is called a *control configuration* at t. A *configuration change* takes place when a control action changes one or several actuator outputs.

An actuator may have an embedded control system which secures the achievement of the intended actuator output within a specified range and time interval. However, for the object PC which executes the control action, the embedded control system of the actuator is not relevant. What is relevant for the object PC is the relation between its control actions and the corresponding actuator outputs.

We shall assume in this book that a control action of an object PC **always** realizes the intended actuator outputs it is meant to result in. If the actuator does not realize its output as intended, this is a control problem at the level of the actuator, but not at the level of the object PC which has this actuator in its actuator interface.

The control of the plant is the result of the interaction between the current state of the plant and the current actuator output. In the example from **Figure 18**, the valve angle may determine the flow level of the water in a tank. A sequence of control actions of the object PC result in a sequence of actuator outputs which in turn result in a sequence of corresponding plant outputs.

The effect of a control action upon the controlled plant is realized in two stages as follows.

Let us assume that the state interpreted to true at t is $S_i=(y_i,\ u_{i,j})$, that is, the control action $u_{i,j}$ is executed at t, given that the plant output y_i is interpreted to true at t. During stage *1*, the control action $u_{i,j}$ realizes the intended actuator output, or in other words realizes a configuration change. In the example from **Figure 18**, the angle of the valve may change from 0° (fully closed) to 30°.

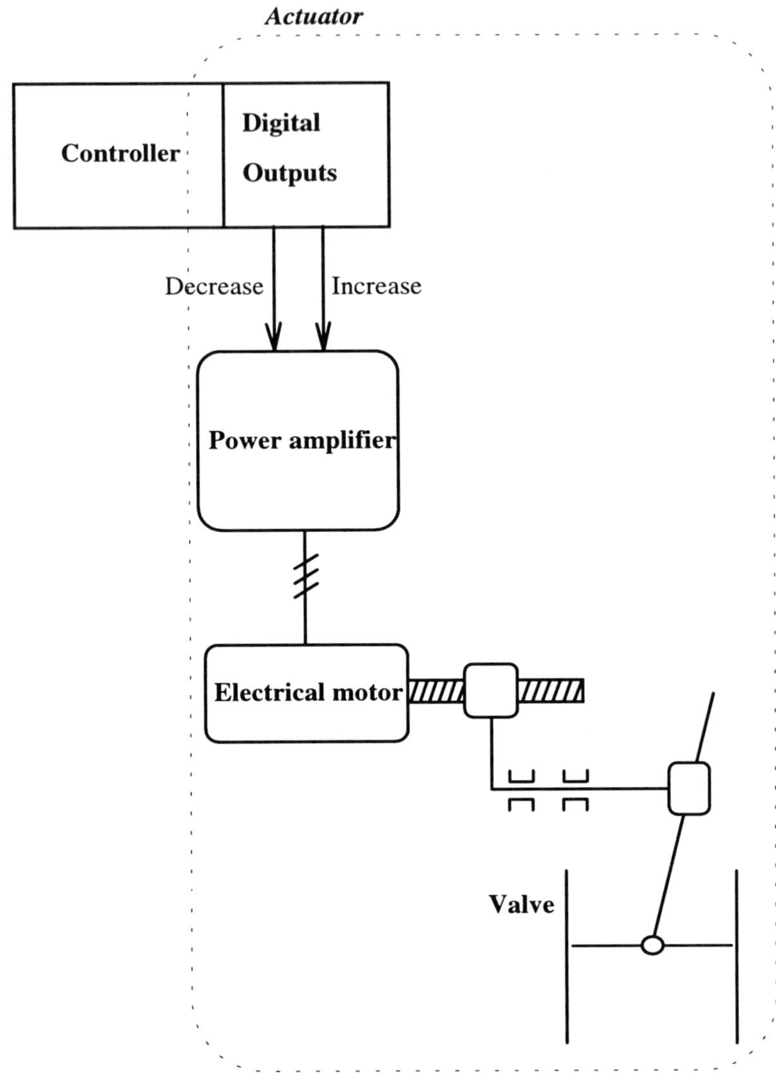

Figure 18 Actuator example.

The realization of the actuator output is not in general instantaneous, but occurs within a certain time interval $[t, t+n]$ $(n \geq 1)$. During this time interval the plant formula y_i describing the plant output at t interprets to true at each sampling instant $t+1$, $t+2$, ...,$t+n-1$ and finally, the intended actuator output is realized at $t+n$. In our example we will have that the final 30° angle is achieved at $t+n$ $(n \geq 1)$. As a result of the configuration change at $t+n$, stage 2 begins. In this stage, the plant output y_i

changes to a new plant output y_j at exactly the same time when the configuration change takes place, i.e., at $t+n$. That is, at $t+n$ the plant formula to interpret to true is y_j that is, $y_j(t')=1$. In our example, the formula y_j may interpret to true when the water level in the tank increases over a pre-specified value. Meanwhile, up to $t+n-1$ (i.e., from t till $t+n-1$), the plant output which continues to interpret to true is y_i.

In this context, when describing control with actuators one has to be aware of the following general characteristics of this type of control:

- **Time delays**: The control action $u_{i,j}$ acts upon the plant by means of a dynamical system, namely the actuator. The two stages described above (an actuator output in the process of a change and no change in previous plant output, followed by a simultaneous change in actuator and plant outputs), occur such that it may take a variable-length time interval until the expected plant output y_j is interpreted to true. This time interval has two components: a time interval $[t,\ t+n-1]$ during which the actuator output is in the process of a change (actuator dynamics) followed by a time interval $(t+n-1,\ t+n]$ during which the actuator and plant outputs change (plant dynamics).

- **External actions.** One and the same actuator may have different effects upon the plant, depending on how expected or unexpected external actions act upon the plant while the actuator output is in the process of a change, i.e., during the time interval $[t,\ t+n-1]$. For example, a water tank may have two pipes: an input pipe for filling the tank with water and an output pipe for water consumption. The input pipe has an actuator (such as a valve), which controls the flow of the liquid in the tank. The water flow through the output pipe depends on the water consumption and cannot be controlled (an external action). If the actuator of the input pipe requires 5 minutes to open fully, then the level of the water in the tank 5 minutes after the execution of the *open* control action depends not only on the water filled in the tank after the open command was executed but also on the water that was removed from the tank through the output pipe due to random water consumption.

- **Assumptions about actuator and plant outputs.** An object PC acts always under a number of assumptions with respect to e interaction between its actuators and the plant. The actuator outputs change always in pre-defined ways following the execution of a control action. No surprises, or other alternative possibilities to this change are feasible. This implies that a control action **always** achieves the configuration change it is intended to realize. However during the second stage described above, a control action relies always on assumptions about the effects of the actuator output upon the plant: for a state transition $S_i \rightarrow S_j$ the post condition y_j of the state S_i materializes at $t+n$ only if the interaction between the actuator output and the plant output y_i at t turns out to be as expected. In other words, we consider an object PC for which it is absolutely certain that control actions always realize intended configuration changes, but there is **a degree of uncertainty** as to whether the plant would react in some pre-specified manner to actuator outputs.

- **Actuator hierarchy.** Often an actuator consists of several underlying physical devices. Therefore a control action $u_{i,j}$ may act through a chain of devices until the actuator as a whole changes its output. The fact that an actuator changes its output as specified is often ensured by a control system built into the actuator. Therefore the distinction about what is an actuator and what is a controller in a hierarchical control system depends on the hierarchical level of the object PC: normally, all the sub-systems between an object PC and the plant under its direct control may be considered as the *actuator* of that particular object PC. Therefore an actuator can be a whole plant that has its output acting upon some other plant (the latter being the plant under control), a servomechanism, or a simple device such as a breaker. However, for each object PC it can be uniquely determined which is the actuator for this particular object PC. An actuator can be recognized as any device or system which has the relevant outputs for the plant under control and these outputs can be realized solely by control actions. In other words, external actions cannot change the output of an actuator, only control actions can.

Furthermore, the need for explicit representation of the actuator interface of the object PC can be motivated as follows. In applications (and as described in Chapter 3), a plant formula represents both plant outputs and actuator outputs. The interpreter program of an object PC interprets plant formulas at each sampling instant. If the current materialized state is $(y_i, u_{i,j})$, the same plant formula y_i may be interpreted to true a number of times until the actuator and the plant realize their intended outputs. However, it is not only the control action $u_{i,j}$ which when executed can lead to the materialization of the particular plant output y_j of the consecutive state S_j. External actions may occur as well and they may have the same effect on the plant as the effect of the control action $u_{i,j}$. Hence, if the plant formula y_j of the consecutive state S_j represents only plant outputs and not the actuator outputs relevant for its materialization, it may happen that due to the occurrence of some external action, y_j is interpreted to true while the actuator has not yet realized its intended output. To avoid this type of ambiguous situation, the plant formula y_j includes a pair of subformulas which are interpreted to true when both the actuator and the plant have realized their intended output.

To illustrate the points above let us consider an example by Ravn et. al. in their work on the specification and verification of requirements of real-time systems using formulas in duration calculus [1]. The example described in [1] consists of a computer-controlled gas burner. This control application is safety-critical since an accident may occur if an excessive amount of unburned gas leaks into the environment. Small gas leaks cannot be avoided during ignition. A burning flame may also be blown out causing some gas to leak before the failure is detected. The gas burner is controlled by a thermostat and the gas is ignited by an ignition transformer, as shown in .

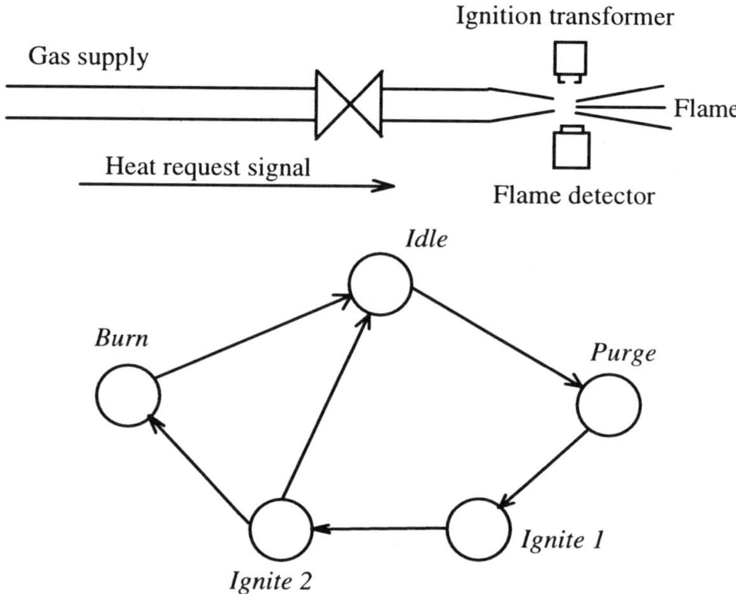

Figure 19 Gas burner [1].

The informal requirements for the object PC are the following:

1. For safety reasons, the gas must never leak for more than *4* seconds in any period of at most *30* seconds.

2. *Heat request off* shall result in a flame being *off* after *60* seconds.

3. *Heat request on* shall, after *60* seconds, result in gas burning unless an ignite of flame failure has occurred. (An *ignition failure* happens when the gas does not ignite after *0.5* seconds. The *flame failure* happens when the flame disappears while gas is supplied).

The gas object PC has the following states:

    Idle: Await heat request; no gas and ignition. The controller enters the *Purge* state on *Heat request*.

Purge: Pauses for *30* seconds and then it entered *Ignite 1* state.

Ignite 1: Starts ignition and gas supply and enters after 1 second the *Ignite 2* state

Ignite 2: Monitors the flame and enters the *Burn* state

Burn: Ignition is switched off, but the gas is still supplied. The *Burn* phase is stable until heat request goes off. The gas is then turned off and the *Idle* state is entered.

If a flame is not sensed within 1 second in the *Ignite 2* state, i.e., there is an *ignition failure*, or if the flame disappears during the *Burn* state (*flame failure*), then the gas is turned off and the *Idle* state is entered. The *30* seconds *Purge* pause ensures a sufficient interval between periods of leaking gas.

This example shows a case when the border between what is a *control action* and what is an *expected external action* is narrow. The control goal in this example is to ensure that in no case is the gas leaking without burning more than a specified time interval. The actuator in this example is the ignition transformer. During the state *Ignite 2*, the gas supply and the ignition transformer are turned on, however the time when the flame appears is not known. Moreover several ignitions may occur as long as the gas total leaking time is under the specified limit and the flame is present for less than 1 second. For the control state that monitors the gas leakage time, the existence of a flame, detected by the flame detector, has the appearance of an expected external action, although is in fact a control action. The control action to turn on the ignition transformer, is done under the assumption that a specific plant output, i.e., the flame, will occur at some future time. Although the transformer is assumed to initiate an ignition, the precise time when the ignition results in a flame is not known. The expected external action in this case is that the gas leaks for more than the specified time interval. Yet, this expected external action is an indirect result of the controller action to open the gas valve. For this example, the explicit representation of the actuator output, i.e., that the ignition transformer is on, is essential. If the state of the ignition transformer, (i.e., the actuator output) is not represented, then each time the controller senses that the flame disappears, it cannot detect if it is as an ignition failure of the type that requires another try (if the gas leakage is within limits), or if it is a flame failure, in which case the controller should enter the idle state.

Some examples of actuators and actuator outputs are the following:

- *Electrical breaker*: the relevant actuator output is the ON or OFF position of the breaker.

- *Industrial robots*: the relevant actuator output is the application program of the robot in applications where a superior control system can change the application program in each of a number of robots in a cell, (factory automation application).

- *Electrical generators*: A number of electrical generators can deliver electrical wer into a power network. Each generator has its own object PC, however each generator together with its PC is only an actuator. The relevant actuator output is then the electrical power a generator delivers. A main object PC performs control actions such as starting, stopping and loading generators. The controlled variable is the electrical energy delivered by the system of generators into the power network.

Relative to the three types of control knowledge from Section 3.3, the control configuration belongs to the ontological type of control knowledge: it further specializes the relation between plant outputs and control actions (via actuators) at each t.

4.1.2.   Control Configuration Representation

Assume n actuator outputs $\{x_1^a,\ldots,x_n^a\} \subset X$. These actuator outputs are a subset of the plant signals X where x_i^a takes its value in the physical domain X_i^a. Each x_i^a has assigned at a sampling instant t a value from X_i^a. This value is denoted by $x_i^a(t)$ and is called the *interpreted value* of x_i^a at t.

A *constrained actuator output*, or an *actuator constraint* on x_i^a, denoted as v_i^a, is a binary relation between x_i^a and certain physical value or a function of physical values. The set of actuator constraints associated with x_i^a is $V_i^a = \{v_{i,1}^a,\ldots,v_{i,m_i}^a\}, i = 1,\ldots,n$. Each actuator constraint is interpreted at t as follows:

$$v_{i,j}^a(t) = 1 \text{ if } v_{i,j}^a \text{ holds at } t \text{ for } x_i^a(t)$$
$$v_{i,j}^a(t) = 0 \text{ if } v_{i,j}^a \text{ does not hold at } t \text{ for } x_i^a(t)$$

A *configuration formula* is a conjunction of the form:

$$c^p = v_{1,k_1}^a \wedge v_{2,k_2}^a \wedge \ldots \wedge v_{n,k_n}^a$$

where $k_i \in \{1,\ldots,m_i\}$. (The upper index of v refers to a specific actuator output and the lower index refers to a specific actuator constraint for this output. The integer m_i denotes the maximum number of constraints for the actuator output with index i).

The *set of configuration formulas* is $C = \{c^1, c^2, \ldots, c^l\}$, where for each index i, j, we have that $c^i \not\sqsubset c^j$ (i.e., c^i is not subsumed by c^j). A configuration formula interprets to true, $c^i(t) = 1$, iff each conjunct in it interprets to 1 at t. A configuration formula interprets to false, $c^i(t) = 0$, iff at least one conjunct interprets to 0. At each t there is exactly **one** configuration formula that interprets to true.

Example.

Suppose we have two electrical breakers associated with a PC. The outputs of the two breakers are x_1^a and x_2^a where the physical domains are $x_1^a = \{closed, open\}$, $x_2^a = \{closed, open\}$. Thus at time t we may have: $x_1^a(t) = open$ and $x_2^a(t) = open$, that is, both x_1^a and x_2^a are interpreted to *open* at t. The set of actuator constraints associated with each actuator can be like:

$$V_1^a = \{v_{1,1}^a, v_{1,2}^a\} \text{ where } v_{1,1}^a \equiv (x_1^a = open), v_{1,2}^a \equiv (x_1^a = closed)$$
$$V_2^a = \{v_{2,1}^a, v_{2,2}^a\} \text{ where } v_{2,1}^a \equiv (x_2^a = open), v_{2,2}^a \equiv (x_2^a = closed)$$

Thus for $x_1^a(t) = open$ and $x_2^a(t) = open$ we will have:

$$v_{1,1}^a(t) = 1, v_{1,2}^a(t) = 0, v_{2,1}^a(t) = 1, v_{2,2}^a(t) = 0$$

A configuration formula can be:

$$c^1 = v_{1,1}^a \wedge \neg v_{2,1}^a$$

and thus, we will have that:

$$c^1(t) = v_{1,1}^a(t) \wedge \neg v_{2,1}^a(t) = 1 \wedge 0 = 0$$

i.e., the configuration formula c^1 is interpreted to false at t.

4.1.3.   Configuration Formulas And Proper Plant Formulas

As shown in Chapter 3, Section 3.1.2, the plant signals in the set $X = \{x_1, x_2, \ldots\}$ are arguments for constrained plant signals v_i. The plant formulas are then represented in terms of boolean functions of constraint plant signals. In Chapter 3, the actuator outputs are not distinguished by special notations in the representation of the plant signals.

The actuator signals in the subset $X^a = \{x_1^a, \ldots, x_n^a\}$ of X are used in Section 4.2. of this chapter as arguments for constrained actuator outputs v_i^a and control configurations are then represented in terms of conjunctions of constrained actuator outputs. However, whether or not control configurations are explicitly represented, this does not affect the way the pre- and post-condition of each control action are interpreted. This is so since the ontological and operational knowledge are the same in both cases. In other words, when a plant formula y_i interprets to true at t, then the corresponding configuration formula c^p that has realized this plant output must interpreted to true as well. That means c^p is a subformula of y_i characterizing exclusively actuator outputs. The remaining part of y_i (i.e., which is not c^p), characterizes the plant outputs alone.

Thus we can re-write a plant formula y_i as the pair (z_i, c_i^p) where:

- c_i^p is the p-th *configuration formula* and c_i^p is a subformula of y_i.

- z_i is called the *proper plant formula* and represents the plant output alone and which plant output is due to the control configuration represented by c^p. The proper plant formula z_i is also a subformula of y_i such that $y_i = z_i \wedge c_i^p$

We assume that a plant formula y_i can be always separated into the pair (z_i, c_i^p) such that:

$$y_i(t) = 1 \ \textit{iff} \ z_i(t) = 1 \ \textit{and} \ c_i^p(t) = 1$$

4.2 Actions

4.2.1.   Control actions

Let us assume that at time t the control configuration is c_i^p, i.e., $c_i^p(t)=1$. When the object controller executes a control action $u_{i,j}$ at t, it changes the actuator output represented by c_i^p to a new actuator output c_j^q. Therefore we shall use the notation $u_{i,j}^{p,q}$ for a control action. The PC reads the new actuator output signals x_i^a and determines which actuator constraints $v_{i,k}^a$ are interpreted to true. Furthermore, by using the actuator constraints that are interpreted to true it determines the configuration formula that is interpreted to true. In this way, it determines a new configuration formula c_j^q which is different from the old configuration formula c_i^p. Therefore, a control action changes the control configuration (in other words a control action makes a *re-configuration*).

From the above we see that each control action is mapped to two different configuration formulas. If the control action $u_{i,j}^{p,q}$ is mapped to the configuration pair (c_i^p, c_j^q) and if:

- the configuration c_i^p is interpreted to true at time t, $c_i^p(t)=1$, and

- the control action $u_{i,j}^{p,q}$ is executed at t, $u_{i,j}^{p,q}(t)=1$,

then the configuration formula c_j^q will **certainly** interpret to true at some later time $t+n$, $n \geq 1$, i.e., $c_j^q(t+n)=1$. The length of the time delay from t to $t+n$, depends on the dynamic properties of the actuator.

The actuator output realized after the execution of the action $u_{i,j}^{p,q}$, interacts with the plant. This interaction results in some predefined plant output. As already described in Chapter 3, Section 3.3.1, the ontological knowledge specifies the interaction between the object PC and the plant, that is, the interaction between actuator outputs and plant outputs.

The state transitions of a PC are specified by the operational knowledge (see Chapter 3, Section 3.3.2). Since each state transition results in a pre-specified plant output and since the ontological knowledge gives the actuator output for each corresponding plant output, it follows that the operational knowledge determines also the mapping between control actions $u_{i,j}^{p,q}$ and configuration pairs (c_i^p, c_j^q).

From the above, it can be seen that the association between the control actions $u_{i,j}^{p,q}$ and pairs of configurations (c_i^p, c_j^q) is determined solely by the control knowledge (i.e., the operational and ontological knowledge). Since the control knowledge underlying the state set S is constant, it follows that when a control action $u_{i,j}^{p,q}$ performs a configuration change, the new configuration that interprets to true at $t+n$ depends only on the previous configuration c_i^p that was interpreted to true at t and on the control action $u_{i,j}^{p,q}$ that was executed at t.

From above, the relevant observation for ontological control is that a change of the control configuration does not depend on expected or unexpected external actions that occur in the plant. Only control actions can change control configurations.

## 4.2.2.	External actions

In the following sections, we use the expression "external actions" to denote both expected and unexpected external actions. When necessary, we state explicitly the distinction between them.

External actions affect the plant output. External actions may be considered as changes of the plant output caused by actuators which do not belong to the object PC. Since these changes are not due to the object PC, the control configuration remains unchanged after the occurrence of an external action. As already defined in Section 3.1.6, an external action may occur at any time and it changes the plant output in a pre-determined way if it is an expected external action, or it simply results in an arbitrary plant formula being interpreted to true if it is an unexpected external action. By this definition we exclude accidental events such as destructive actions upon actuators or different disturbances which hinder the normal function of actuators. Ontological control refers to the normal function of an object PC for which the actuators are fully under control and external actions change only plant outputs, but not actuator outputs.

Let us assume that at t the configuration formula is c_i^p and it will remain unchanged till $t+n-1$. Let the proper plant formula that corresponds to the configuration formula c_i^p be z_i. If, during the interval $[t, t+n-1]$ the expected external action $u_{i,a}^{ext}$ occurs at t', the controller configuration c_i^p remains the same, but the proper plant formula changes from z_i to z_a. Therefore we have a new state with $c_i^p(t')=1$ and $z_a(t')=1$. If instead of the expected external action $u_{i,a}^{ext}$, the expected external action $u_{i,b}^{ext}$ occurs at t', then the proper plant formula changes to z_b but the configuration c_i^p remains unchanged again and we have a new state materialized with $c_i^p(t')=1$ and $z_b(t')=1$.

Let the state interpreted to true at t be S_i. We note the following:

- *Expected external actions.* All of the proper plant formulas that are expected to materialize following the occurrence of an expected external action at S_i are known in advance. They are part of states which have the same configuration formula, say c_i^p (i.e., the configuration of S_i), but distinct proper plant formulas. Let the set of all the proper plant formulas of these states be $Z_i'=\{z'_a, z'_b, ..., z'_x\}$.

- *Unexpected external actions*. The plant formula that materializes following the occurrence of an unexpected external action at S_i is not known in advance. However, we know that some plant formula materializes always, following the occurrence of an unexpected external action (see Section 3.1.6). Still, since the configuration formula c_i^p remains unchanged after the occurrence of the unexpected external action, the plant formula that can materialize is one that has its corresponding configuration formula equal to c_i^p. Let the set of all the proper plant formulas z_k that appear in pairs (z_k, c_i^p), with the same configuration formula c_i^p be Z_i. With this notation, the proper plant formula that materializes after the occurrence of an unexpected external action belongs to $Z_i \setminus Z_i'$.

We have assumed that a current plant formula can change to another plant formula only due to the occurrence of one of the following: a control action, an expected external action, or an unexpected external action. For a given control configuration, only expected, or unexpected external actions can change the plant formula so that the given configuration formula is preserved. Therefore, given that the configuration formula c_i^p remains unchanged, the proper plant formula z_l that can materialize at a later time can be any one that results from the occurrence of an expected, or unexpected external action, i.e., $z_l \in (Z_i \setminus Z_i') \cup Z_i'$, that is, $z_l \in Z_i$.

4.3 Well-Defined Controller State With A Configuration

Here we will re-introduce the notion of a well-defined state by augmenting it with a configuration formula.

A well-defined state with a configuration is denoted as:

$$S_i^p = ((z_i, c_i^p), u_{i.j}^{p.q})$$

where:

- z_i is the i-th plant formula

- c_i^p is the p-th configuration formula such that if $c_i^p(t)=1$ then $z_i(t)=1$

- $u_{i,j}^{p,q}$ is a control action with a precondition pair $(z_i,\ c_i^p)$ and a post-condition pair $(z_j,\ c_j^q)$ such that when $z_i(t)=1$, $c_i^p(t)=1$ and $u_{i,j}^{p,q}(t)=1$ it is **certain** that $c_j^q(t+n)=1$ and it is **expected** that $z_j(t+n)=1$. Furthermore we also have that both of c_i^p and z_i are interpreted to true not only at t but also at each sampling instant of the open interval $(t,\ t+n-1)$.

A well-defined state with an expected external action $u_{i,j}^{ext}$ is denoted as:

$$S_i^p=((z_i,\ c_i^p),\ u_{i,j}^{ext})$$

where:

- z_i is the i-th plant formula

- c_i^p is the p-th configuration formula such that if $c_i^p(t)=1$ than $z_i(t)=1$.

- $u_{i,j}^{ext}$ is an expected external action with a precondition pair $(z_i,\ c_i^p)$ and a post-condition pair $(z_j,\ c_j^p)$ such that if $z_i(t)=1$, $c_i^p(t)=1$ and $u_{i,j}^{ext}(t)=1$ then it is **certain** that $c_j^p(t')=1$ and it is **expected** that $z_j(t')=1$ where t' is in $[t,\ t+n)$. (Note that an expected external action does not change the configuration formula c^p. The configuration can change only at $t+n$ due to a control action. Furthermore, the occurrence of an expected external action can only be detected at a time later than its actual occurrence).

With respect to a unexpected external action we again (as in Chapter 3) have no symbolic notation for it . Let us consider the state $S_i=((z_i,\ c_i^p),u_{i,j}^{p,q})$ and let there be no expected external action $u_{i,k}^{ext}$. If at t we have that $z_i(t)=1$, $c_i^p(t)=1$ and $u_{i,j}^{p,q}(t)=1$ and at t' (t' is in $[t,\ t+n)$) we obtain $z_j(t')=0$, $c_i^p(t')=1$ and $z_k(t')=1$, then this is due to an unexpected external action. Note here that a unexpected external action does not change the control configuration. Furthermore, the occurrence of such an action can be detected only at t'. Since there is no symbol corresponding to such an action there is no notation for a state with unexpected external actions.

In what follows we will call a well-defined state with a configuration just a state.

4.3.1 A Consecutive State

The state S_j^q is consecutive to the state S_i^p iff:

$$S_i^p=((z_i,\ c_i^p),\ u_{i,j}^{p,q})\text{ and }S_j^q=((z_j,\ c_j^q),\ u_{j,k}^{q,r})$$

That is, the precondition pair of $u_{j,k}^{q,r}$ is the post-condition pair of the control action $u_{i,j}^{p,q}$.

In this context, suppose that there exists an expected external action $u_{i,j}^{ext}$, with the precondition pair $(z_i,\ c_i^p)$. Since the postcondition pair of $u_{i,j}^{ext}$ is $(z_j,\ c_j^p)$ and thus, its pre- and post-condition pairs have the same configuration formula, the state $S_j^p=((z_j,\ c_j^p),\ u_{j,l}^{p,u})$ is not the consecutive state of either $S_i^p=((z_i,\ c_i^p),\ u_{i,j}^{p,q})$ nor $S_i^p=((z_i,\ c_i^p),\ u_{i,j}^{ext})$.

As in Chapter 3, a material state is a state where both its precondition pair and its control action are interpreted to true. In this case we say that the state is materialized. A partial state is a state whose control action part does not contain a particular control action and such a state is interpreted to true when its precondition part is interpreted to true. In this case we say that a partial state has been materialized. A goal state is a state of the form $S_i^p=((z_i,\ c_i^p),\ u_{i,i}^{p,p})$. The control action $u_{i,i}^{p,p}$ denotes a *null* action such that if at time t $z_i(t)=1$ and $c_i^p(t)=1$ then in the absence of external actions, for each $n>0$ will result $z_i(t+n)=1$ and $c_i^p(t+n)=1$.

The properties of a state set of an object PC follow from the properties of the plant formulas (Section 3.1.5), of the control actions (Section 3.1.6) and from the properties of the configuration formulas (Section 4.1.3):

- Two proper plant formulas may be subsumed: there may exist two proper plant formulas z_i and z_j such that $z_i \subset z_j$.

- Two configuration formulas cannot be subsumed: for each two distinct configuration formulas c^p and c^q, we have that $c^p \not\subset c^q$ (see Section 4.1.2).

- At each time instance t there is exactly one pair $(z_i,\ c_i^p)$ that can interpret to true (or materialize), i.e., $z_i(t)=1 \wedge c_i^p(t)=1$. That means, although the proper plant formulas may be subsumed, there are no two states in S, $((z_i,\ c_i^p),\ \text{-})$ and $((z_j,$

c_j^q), -) such that $z_i \subset z_j$ <u>and</u> $c_i^p \subset c_j^q$. Note that there may exist states for which z_i materializes, but there is always exactly one state for which both z_i and c_i^p materialize at t.

- For each distinct pair (z_i, c_i^p) there exists at least one control action $u_{i,j}^{p,q}$. At each t only one control action can be executed (i.e., interpreted to true).

- If $z_i(t)=0$, $c_i^p(t)=1$ and $u_{i,j}^{p,q}(t)=1$ then $z_j(t+n)=0$ and $c_j^q(t+n)=1$. In other words a control action always realizes its corresponding control configuration, but if the proper plant formula of its precondition pair was not interpreted to true when the action was executed, then the proper plant formula of its post-condition pair cannot materialize.

- The set of configuration formulas C, proper plant formulas Z and control actions U is not modified during control by adding or deleting configuration formulas, proper plant formulas, and/or control actions.

4.3.2 State Transitions

We re-consider here the notions of state transitions and transitions defined in Chapter 3 in the context of states with explicit configuration formulas

4.3.3 State transitions and transitions

A *state transition* takes place from a state to any one of its consecutive states and is denoted as $S_i^p \rightarrow S_j^q$. Thus, a state transition can only be due to control actions.

A *transition* takes place from a state S_i^p to a state S_j^p and is denoted as S_i^p --> S_j^p. Observe that in the case of a transition the configuration formula c^p of the state S_i is preserved in the second state of the transition S_j. This in turn implies that the state S_j^p is not a consecutive state of the state S_i^p. This type of transition is due to external actions (expected and unexpected) and de-synchronizations due to VOA, ill-represented states, etc..

A transition due to external actions (expected or unexpected) takes place due to an action **not** executed by the object PC and thus, such an action cannot change the actuator outputs under the direct control of the object PC. The effect of the external action upon the plant is dependent on the control configuration, but it does not affect the control configuration. Therefore a state transition from S_i^P to S_k^P due to the expected external action $u_{i,k}^{ext}$ is denoted as follows:

$$S_i^P \dashrightarrow S_k^P \equiv ((z_i,\ c_i^p),\ u_{i,k}^{ext}) \dashrightarrow ((z_k,\ c_k^P),\ u_{k,l}^{p,r})$$

A transition from S_i^P to S_k^P due to unexpected external action is denoted as follows:

$$S_i^P \dashrightarrow S_k^P \equiv ((z_i,\ c_i^p),\ u_{i,j}^{p,q}) \dashrightarrow ((z_k,\ c_k^P),\ u_{k,l}^{p,s})$$

As already described in Chapter 3 we do not have symbols for unexpected external actions since these are not defined in advance as expected external actions are. However, if the above transition materializes, one can a posteriori construct a symbol for the unexpected external action which has taken place. In the context of the above transition one can say that a unexpected external action $u_{i,k}^{ext}$ has occurred and furthermore, one can also say that its precondition pair has been $(z_i,\ c_i^p)$ while its post-condition pair has been $(z_k,\ c_k^P)$.

4.3.4 Material state transitions and transitions

A material state transition takes place between a material state and any one of its material consecutive states and is denoted as $S_i^P(t) \rightarrow S_j^q(t+n)$. A *material transition* takes place whenever the states S_i^P and S_j^P materialize consecutively and is denoted as $S_i^P(t) \rightarrow S_j^P(t+n)$.

A material state transition takes place in two stages (see Section 4.1.1): first the control action changes the control configuration and second, the new control configuration results in a simultaneous change of the plant output.

Let a state transition $S_i^P \rightarrow S_j^q$ be given as,

$$((z_i,\ c_i^p),\ u_{i,j}^{p,q}) \rightarrow ((z_j,\ c_j^q),\ u_{j,k}^{q,r})$$

Let us assume that at time t the state S_i^P materializes. This means that the proper plant formula z_i is materialized at t, $z_i(t)=1$; the configuration formula c_i^p is also materialized at t, $c_i^p(t)=1$; the control action $u_{i,j}^{p,q}$ is executed at t, $u_{i,j}^{p,q}(t)=1$. Then at

$t+n$ $(n\geq1)$ the post-condition $(z_j,\ c_j^q)$ interprets to true and the control action $u_{j,k}^{q,r}$ is executed at this same time. Up to $t+n$ it is the precondition $(z_i,\ c_i^p)$ which interprets to true at each sampling instant of the interval $[t,\ t+n-1]$. During this time interval, both of c_i^p and z_i are in the process of a change towards z_j and c_j^q respectively. In what follows we shall consider several cases of a material state transition.

Materializing a state transition with no preceding inner state transitions.

Let at t the state $S_i^p = ((z_i,\ c_i^p), u_{i,j}^{p,q})$ be materialized. At the next time instance, $t+1$, the post-condition pair $(z_j,\ c_j^q)$ of $u_{i,j}^{p,q}$ materializes: i.e., both the proper plant formula z_j and the configuration formula c_j^q are interpreted to true after exactly one sampling instant.

A material state transition which takes place in the time interval $[t,\ t+1]$ is called a *transient transition*. The state from which such a transition is initiated is called a *transient state*. Examples of such state transitions involve electrical breakers, valves, or other devices which produce swift changes of their output.

Materializing a state transition with preceding inner state transitions.

At all sampling instances in the interval $[t,\ t+n-1]$, the precondition $(z_i,\ c_i^p)$ of $u_{i,j}^{p,q}$ interprets to true. At each time after t when the precondition $(z_i,\ c_i^p)$ of $u_{i,j}^{p,q}$ interprets to true, the state S_i^p is said to be in an *inner state transition*. Inner state transitions happen because the configuration c_j^q of the expected state S_j^q cannot materialize in one sampling instant due to the dynamics of the actuators and therefore, the proper plant formula z_j cannot materialize either. This case is characteristic for slow, continuous actuators.

Materializing state transitions due to external actions

As for case 2, let us assume that the state $S_i^p = ((z_i,\ c_i^p), u_{i,j}^{p,q})$ is in inner state transitions during the interval $[t,\ t+n-1]$ (**Figure 20**). This means, that in the absence of any external action, the precondition pair of $u_{i,j}^{p,q}$, (i.e., $(z_i,\ c_i^p)$) materializes at any sampling instant during a time interval $[t,\ t+n-1]$ and the post-condition pair of $u_{i,j}^{p,q}$ does not materialize during the same time interval. However, suppose that at some time t' preceding the time $t+n-1$, an external action (expected or unexpected) changes the plant output. This has the effect that the proper plant formula z_i of the state S_i is interpreted to false at t', but the proper plant formula z_k of some state $S_k^p = ((z_k, c_k^p), u_{k,n}^{p,s})$ materializes at t'. The configuration formula at t' is still c_i^p since it can only change to c_j^q at $t+n$. The proper plant formula z_i of the state S_i^p cannot materialize any more since the state $S_k^p = ((z_k, c_k^p), u_{k,n}^{p,s})$ materializes at t' and thus, the control action $u_{k,n}^{p,s}$ is executed at this time which will start changing the configuration formula from c_k^p to c_n^s rather than from c_k^p to c_j^q. Since only the configuration c_i^p can bring about the materialization of the plant formula z_j and we have instead the configuration c_n^s the plant formula z_j cannot materialize any more.

If the state S_i^p is instead a transient state it is obvious that an external action can prevent the expected consecutive state S_j^q to materialize at $t+1$ only if the external action can change the plant output in a time less than one sampling instant. Otherwise the effect of the occurrence of an external action will take place after the consecutive state of S_i^p, that is S_j^q, has materialized at $t+1$. In practice, the second scenario is normally the case.

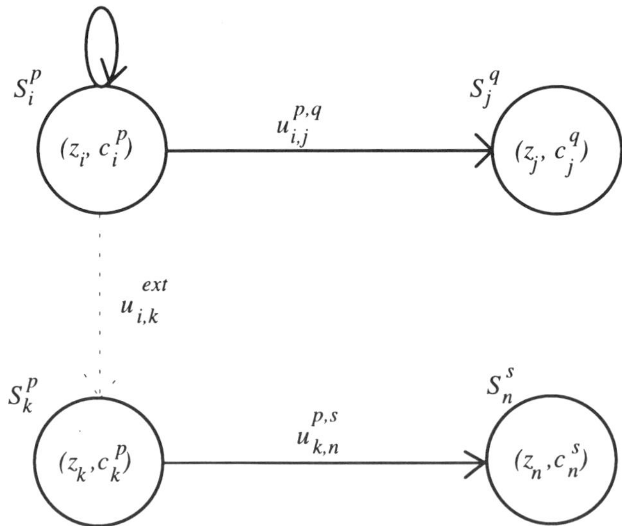

Figure 20 State transition due to external action.

4.5 De-Synchronization

Let us assume that the state $S_i^p=((z_i,\ c_i^p),u_{i,j}^{p,q})$ is materialized at time t. Let us also assume that the state expected to materialize at $t+n$ ($n \geq 1$) is $S_i^q=((z_j,\ c_j^q),\ u_{i,j}^{q,v})$. In analogy to the case of de-synchronization from Chapter 3, a de-synchronization takes place when at $t+n$ the materialized post-condition pair is different from $(z_j,\ c_j^q)$ and thus, the state with this post-condition pair is not amongst the consecutive states of S_i^p. For example, the proper plant formula materialized at $t+n$ is say, z_k rather than z_j. Thus, the state that can be materialized at $t+n$ is not the consecutive state of S_i^p since any of the consecutive states of S_i^p has z_j in its precondition pair rather than z_k.

4.6 A Well-Determined State Set

In this section we formulate a number of constraints on the elements of the state set S and thus, obtain the so called *well-determined state set*. A well-determined state set contains states that have restricted allowable transitions. These restrictions, in turn, eliminate the presence of non-optimal goal paths. Using well-determined state sets, we can show in Chapter 5 that a de-synchronization due to VOA can be distinguished from de-synchronizations due to unexpected external actions and ill-represented states. The constraints on the elements of the state set are encountered in practice in some carefully designed application programs of object PCs, but so far there has been no explanation for why these constraints should be obeyed and how can they be used to distinguish between the causes for a de-synchronization.

4.5.1 The Integrity Property Of S

The state set S contains the state

$$S_i^p=((z_i,\ c_i^p),u_{i,j}^{p,q})$$

or the state,

$$S_k^q=((z_k,\ c_k^q),\ u_{k,j}^{ext})$$

but not both states.

This property implies that there is no state in S that is reachable by both a control and by an expected external action. This is so because the state S_i^p is the first state in a state transition say $((z_i, c_i^p), u_{i,j}^{p,q}) \rightarrow ((z_j, c_j^q), u_{j,m}^{q,r})$ while the state S_k^q is the first state in a transition say $((z_k, c_k^q), u_{k,j}^{ext}) \rightarrow ((z_j, c_j^q), u_{j,m}^{q,r})$. If only one of these two states is allowed to be in S this implies that only one of these two transitions can exist.

This property of S follows from the following control considerations:

- An external action cannot materialize the same plant output as a control action. In other words, if the external actions are performed by actuators not in the actuator interface of an object PC these actuators cannot be used as a substitute for the actuators of the object PC. Thus, the actuators of the object PC are fully under its control and external actions cannot change their output.

- Reciprocally, a control action cannot result in the same plant outputs as an external action. If the external actions are performed by actuators which do not belong to the actuator interface of the object PC the object PC cannot execute them and thus, they cannot be used as a substitute for its control actions.

4.5.2 The Specificity Of Control Configuration Property

Let $S_i^p = ((z_i, c_i^p), u_{i,j}^{p,q})$ be the consecutive state of a state with a control action. Then, there is no other state S_l^p ($l \neq i$) that is the consecutive of a state with a control action and such that S_l has the precondition (z_l, c_l^p).

The specificity of control configuration property implies that if the configuration formula c_i^p belongs to a state S_i^p which is the second state in a state transition due to a control action, then there is no other state that has the same configuration formula c_i^p, and is also the second state in a state transition due to a control action. Thus, another state with the same configuration formula may belong to a state which is the second state in a state transition due to an expected external action, or unexpected external action. The above property does not hold for the initial states in goal paths since these cannot be consecutive states.

The property of specificity of control configuration is based on the following control considerations:

Let us assume that there is a state S_l^t which is the consecutive state of a state with control action (shown in Figure 21) and let the configuration formula of S_l^t be identical to the configuration formula of S_j^q but their proper plant formulas are different: $z_j \neq z_l$ and $c_j^q \neq c_l^t$.

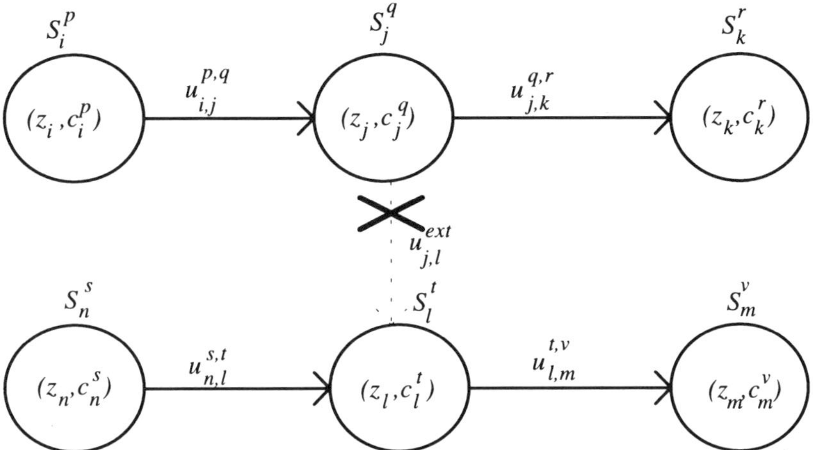

Figure 21 Non-specific control configuration.

The states S_j^q and S_l^t are both consecutive to two states S_i^p, respectively S_n^s such that S_i^p and S_n^s are states with control actions $u_{i,j}^{p,q}$ and $u_{n,l}^{s,t}$. Due to the integrity property there cannot be any $u_{j,l}^{ext}$ such that $((z_j, c_j^q), u_{j,l}^{ext}) \to S_l^t$ (shown by a crossed transition in the Figure 21).

Now according to the above, the states S_j^q and S_l^q have the same configuration, c^q, but there is no external action $u_{j,l}^{ext}$ from S_j^q to S_l^q. This means under one and the same control configuration, the corresponding plant outputs are different. Since there is no external action which can perform a transition from z_j to z_l, this means that the difference in the plant output is due to a particular past control action(s) (or a particular relative order of control actions) which is/are different in the past state transitions leading to S_j^p compared to the past state transitions leading S_l^t. However, what ever these differences may be, they should be represented by distinct control configurations.

4.5.3 Specificity Of Control Action Property

Let $S_t^p=((z_i, c_i^p), u_{i,j}^{p,v})$ be a state. There is no other state S_t^p such that $S_t^p=((z_i, c_i^v),$ $u_{i,j}^{p,w})$ $(v{\neq}w)$.

It is easily seen that the above property allows the existence of only one of the following two state transitions:

$$((z_i, c_i^p), u_{i,j}^{p,v}) \rightarrow ((z_j, c_j^v), u_{j,m}^{v,q})$$

or the state transition,

$$((z_i, c_i^p), u_{i,k}^{p,w}) \rightarrow ((z_k, c_k^w), u_{k,n}^{w,r})$$

The control considerations for the specificity of control action property are as follows. Consider an object PC without a GSO, or with a GSO, but with an incomplete priority order on goal states. Let two states have one and the same precondition pair, but distinct control actions associated with this precondition pair. If the precondition pair materialize at t, then there are no criteria available to select between the two different control actions.

4.5.4 Syntactically Complete State Set

Let S be a state set, Z be the set of proper plant formulas of S, and U^{ext} be the set of expected external actions.

The state set S is *syntactically complete* iff:

- For each proper plant formula $z_i \in Z$ there exists at least one configuration formula c_i^p and a control action $u_{i,j}^{p,q}$ in some states of S and

- For each expected external action $u_{i,j}^{ext} \in U^{ext}$ there exists exactly two proper plant formulas z_i and z_j in some states of S, and

- For each control action $u_{i,j}^{p,q}$ there exist at least one plant formula z_j in some states of S.

The control considerations for this property of the state set are that each plant output should be realisable via a certain actuator output. In its turn, each plant output should be realisable via the execution of a control action. Reciprocally, the execution of each control action should result in a particular plant output via the realization of a particular control configuration. Furthermore, any expected external action should be possible to detect via its effects on the plant outputs and the plant outputs which make its occurrence possible.

4.5.5 A Well-Determined State Set And Its Properties

In this section we restrict a state set to a state set with the properties described in Sections 4.5.1 to 4.5.4. The so restricted state set is called a *well-determined state set*. The restrictions on its elements induce certain properties with respect to consecutive states of a state. These in turn restrict the type of allowable goal paths and induce certain properties for the allowed goal paths.

A Well-Determined State Set

A state set S is *well-determined* iff S has the following properties:

- S has the control integrity property (Section *4.5.1*)

- S has the specificity of control configuration property (Section *4.5.2*)

- S has the specificity of control action property (Section *4.5.3*)

- S is syntactically complete (Section *4.5.4*)

Proposition 4:1 A well-determined state set S has the following properties:

1. *each state $S_i \in S$ has a consecutive state in S*

2. *each state S_m which does not belong to S does not have a consecutive state in S*

Proof.

(1) Let the state $S_i^p = ((z_i,\ c_i^p),\ u_{i,j}^{p,q})$ be an arbitrary state in S. Since S is syntactically complete, then for the control action $u_{i,j}^{p,q}$ there is a z_j in some state of

S, Furthermore, since z_j is in S then there is a c_j^q and a $u_{j,n}^{q,v}$ in some state of S. Since the states in S are well-defined then the state in S that has z_j, c_j^q, and $u_{j,n}^{q,v}$ is $((z_j, c_j^q), u_{j,n}^{q,v})$ which by definition is the consecutive state of S_i^p.

(2) Let the state $S_i^p = ((z_i, c_i^p), u_{i,j}^{p,q}) \notin S$. By the definition of a consecutive state, all states in the set of consecutive states of S_i^p have the same precondition pair (z_j, c_j^q). Furthermore, since S is syntactically complete the set of consecutive states of S_i^p cannot be empty. Since S is well-determined all control actions that have (z_i, c_i^p) as its precondition pair and are of the form $u_{i,j}^{p,v}$ $(q{\neq}v)$ are in S, that is the states $((z_i, c_i^p), u_{i,j}^{p,v})$ are in S. Otherwise, if both $u_{i,j}^{p,q}$ and $u_{i,j}^{p,v}$ are in some states of S and have the same precondition pair this would violate the action specificity property of S. Now consider the consecutive states of a state in S such as $((z_i, c_i^p), u_{i,j}^{p,v})$. First its consecutive states are in S according to (1) of this proposition. Second, any one of its consecutive states, by the definition of a consecutive state, has the same precondition pair (z_j, c_j^v). Obviously this precondition pair is different from the precondition pair (z_j, c_j^q) of $S_i^p \notin S$. Thus, if a state not in S and a state in S have the same precondition pair (i.e., (z_i, c_i^p)) then the precondition pairs of their consecutive states are different (in their configuration formula parts) and cannot be both in S: only the precondition pair of the consecutive states of the state in S is in S while the precondition pair of the consecutive states of the state not in S is not in S.
Q.E.D.

The control considerations underlying the above proposition are the following. When the precondition pair (z_m, c_m^x) is in some state of S but the full state $((z_m, c_m^x), u_{m,i}^{x,p})$ is not in S, this means the control action $u_{m,i}^{x,p}$, when executed under the preconditions (z_m, c_m^x) does not result in a state on a goal path since all the possible ways to materialize this goal path have already been represented in S.

Proposition 4:2 *Each state in a well-determined state S has only one consecutive state.*

Proof.

Let the state $S_i^p = ((z_i, c_i^p), u_{i,j}^{p,q})$ be in S where S is well-determined. Suppose that S_i^p has two consecutive states, that is:

$$((z_j, c_j^q), u_{j,r}^{q,v})$$

and

$$((z_j, c_j^q), u_{j,}^{q,w})$$

In other words, both of the above states have the same precondition pair and different control actions associated with this precondition pair. This is not possible since S is well-determined and the action specificity property which then holds for the elements of S forbids that both of the above states are in S. Thus, only one of them can be in S.

Q.E.D

4.7 Goal Paths In A Well-Determined State Set

In Chapter 3 the notion of a goal path was defined as a sequence of state transitions from an initial state to a goal state. Furthermore, each initial state and goal state had only one optimal goal path connecting them. However, it was also possible that there was more than one non-optimal goal path connecting an initial and a goal state. It was allowed for one and the same state to appear in more than one goal path. That resulted in the existence of non-optimal goal paths and cycles involving state transition sequences from different goal paths. Finally, it was also allowed that one and the same plant formula was in more than one state, these states being on the same optimal goal path and/or on different optimal goal paths. In this section we will show that when the state set is a well-determined one, the restrictions upon its elements make impossible the existence of non-optimal goal paths and of certain type of cycles.

Proposition 4:3. *Let S be a well-determined state set and $S_i^p \in S$. If S_i^p is at least in two places in the same optimal goal path to a goal state S_g then the goal state S_g is not reachable.*

Proof.
Consider the following goal path constructed from the elements of a well-determined state set S:

$$S_1^{p_1} \to S_2^{p_2} \to \ldots \to S_i^{p_i} \to S_{i+1}^{p_{i+1}} \to S_{i+n}^{p_{i+n}} \to S_{i+n+1}^{p_{i+n+1}} \to \ldots \to S_g$$

and let $S_i^p = S_{i+n+1}^{p_{i+n+1}}$. Since each state in S has a unique consecutive state (*Proposition 4:2*), then the consecutive state of $S_{i+n+1}^{p_{i+n+1}}$ will always be $S_{i+n+2}^{p_{i+n+2}} = S_{i+1}^{p_{i+1}}$. The consecutive state of $S_{i+n+2}^{p_{i+n+2}}$ will always be $S_{i+n+3}^{p_{i+n+3}} = S_{i+2}^{p_{i+2}}$, etc. Thus for each state $S_{i+n+k}^{p_{i+n+k}}$, ($k>1$) the only states reachable from it are the states from S_i^p to $S_{i+n+1}^{p_{i+n+1}}$, but not the goal state S_g.

Q.E.D.

This is the reason for requiring the following property of goal paths in the case when the state set from which they are constructed is well determined:

Property 4:1 *Each state in S appears only once in the same optimal goal path.*

The next property of goal paths follows directly from the specificity of control action for a well-determined state set S:

Property 4:2. *Any precondition pair* $(z_i,\ c_i^p)$ *appears only once in some state of one and the same optimal goal path.*

If the above precondition pair is encountered in say two different states on the same optimal goal path this implies that in each one of these states it is associated with a different control action. However, the action specificity property forbids this situation.

The effect of the above two properties is that they eliminate the possibility for control cycles (i.e., cycles due to state transitions only) in one and the same optimal goal path. The reason for this is that, as described already in Chapter 3, a cycle may appear in an optimal goal path if the same state is encountered at least twice in it, or if the goal path has at least two states with the same plant formula. This implies the following property for goal paths constructed from the elements of a well-determined state set S.

Property 4:3 *An optimal goal path has no control cycles.*

The following proposition justifies the next property of the goal paths constructed from the elements of a well-determined state set S. This property forbids the existence of non-optimal goal paths.

Proposition 4:4. *Let S be a well-determined state set and let the state S_i^p in S be in two different optimal goal paths. Then all the states in the two optimal goal paths succeeding S_i^p, are identical and lead to the same goal state.*

Proof.

Consider the two optimal goal paths:

$$S_i^{p_i} \to S_{i+1}^{p_{i+1}} \to S_{i+2}^{p_{i+2}} \to S_{i+3}^{p_{i+3}} \to \dots \to S_{i+n}^{p_{i+n}} \to S_g$$
$$S_j^{p_j} \to S_{j+1}^{p_{j+1}} \to S_{j+2}^{p_{j+2}} \to S_{j+3}^{p_{j+3}} \to \dots \to S_{j+n}^{p_{j+n}} \to S'_g$$

and let $S_{i+3}^{p_{i+3}} \equiv S_{j+3}^{p_{j+3}}$. Since each state in S has a unique consecutive state we have that:

$$S_{i+4}^{p_{i+4}} \equiv S_{j+4}^{p_{j+4}} \dots S_{i+n}^{p_{i+n}} \equiv S_{j+n}^{p_{j+n}}$$

or all the states succeeding $S_{i+4}^{p_{i+4}}$ and $S_{j+3}^{p_{j+3}}$ are identical.

Furthermore, let

$$S_g = ((z_{i+n+1},\ c_{i+n+1}^{p_{i+n+1}}),\ u_{i+n+1,\ i+n+1}^{p_{i+n+1}},\ p_{i+n+1})$$

Then, because all the states succeeding the state belonging to the two goal paths are identical, we will have that

$$S_g' = ((z_{i+n+1},\ c_{i+n+1}^{p_{i+n+1}}),\ u_{i+n+1,\ i+n+1}^{p_{i+n+1}},\ p_{i+n+1})$$

Thus both optimal goal paths lead to the same goal state.

Q.E.D.

In other words, if we allow one and the same state to be in more than one optimal goal path we also will allow the situation when there are at least two optimal goal paths leading to the same goal state. However, this will contradict the definition of an optimal goal path, that is a unique state transition sequence from an initial state to a goal state. That is why we require the following property for goal paths constructed from the elements of a well-determined state set S.

Property 4:4 *Each state is in only one optimal goal path.*

As already described in Chapter 3 non-optimal goal paths are made possible when one and the same plant formula (in this case a pair consisting of proper plant and configuration formulas) is common for at least two states on these optimal goal paths and the control actions in these states are different. However, for a well-determined state set, there cannot exist two states with an identical precondition pair and different control actions associated with this precondition pair due to the property of specificity of control action. This in turn implies the next property of goal paths constructed from the elements of a well-determined state set S.

Property 4:5. *There are no non-optimal goal paths.*

In Chapter 3 we explained that an external action (expected and unexpected) serves the role of a bridge between two optimal goal paths. Thus, during the interpretation of a given optimal goal path the occurrence of an external action gave the possibility for continuing the interpretation on a different optimal goal path. In other words, external actions define transitions from one optimal goal path to another one. The following proposition shows that when the goal paths are constructed from the elements of a well-determined state set an external action cannot define a transition to a state on a different goal path .

Proposition 4:5. *For any states $S_i^p = ((z_i,\ c_i^p), u_{i,j}^{p,q})$ and $S_k^r = ((z_k,\ c_k^r),\ u_{k,l}^{r,s})$ which are on two different optimal goal paths and S_k^r is not an initial state there is no state $S_i^p = ((z_i,\ c_i^p),\ u_{i,k}^{ext})$ which belongs to S.*

Proof.
Let us first note here that though we use the notation for a state with an expected external action that is $((z_i,\ c_i^p),\ u_{i,k}^{ext})$ the result that is to be proved holds for unexpected external actions as well. This is so because the non-existing symbol for a unexpected external action can be provided after the occurrence of such an action since then one knows exactly what the pre- and postconditions of this unexpected action were. Now let us proceed with the proof.

Let us assume that the state $S_i^p = ((z_i,\ c_i^p),\ u_{i,k}^{ext})$ is in S. Then let its consecutive state be $S_k^p = ((z_k,\ c_k^p),\ u_{i,j}^{p,v})$, which is not an initial state. The latter implies that S_k^p is then the consecutive state of another state in S, namely the state $S_m^t = ((z_m,\ c_m^t),\ u_{m,k}^{t,p})$. Thus we will have that both states $S_i^p = ((z_i,\ c_i^p),\ u_{i,k}^{ext})$ and $S_m^t = ((z_m,\ c_m^t),\ u_{m,k}^{t,p})$ are in S. However, the simultaneous presence of both this states in S is in contradiction to the integrity property, that is no state in S can be reached from a state with a control action and a state with an external action. Thus, the state S_i^p is not in S.

Q.E.D.

The above proposition forbids the presence of S_i^p (i.e., a state with external action) in S only when its consecutive state is not an initial state. This is so because when the consecutive state S_k^p of $S_i^p = ((z_i,\ c_i^p),\ u_{i,k}^{ext})$ is an initial state it is not the consecutive state of another state in S and thus, the state S_m^t does not exist. If the state S_m^t does not exist then nothing forbids the presence in S of all three states $S_i^p = ((z_i,\ c_i^p),\ u_{i,j}^{p,q})$, $S_i^p = ((z_i,\ c_i^p),\ u_{i,k}^{ext})$, and $S_m^t = ((z_m,\ c_m^t),\ u_{m,k}^{t,p})$, as long as the latter state is an initial state. This in turn implies the following property.

Property 4:6. *Each state with external action* $S_i^p = ((z_i,\ c_i^p),\ u_{i,k}^{ext}) \in S$ *has a consecutive state which is an initial state of some goal path.*

Note that the consecutive state of $S_i^p = ((z_i,\ c_i^p),\ u_{i,k}^{ext})$ can be the initial state of the same goal path to which the state $S_i^p = (z_i,\ c_i^p),\ u_{i,j}^{p,q})$ belongs.

4.8 The Control set and The Collateral Set Of A State

For ontological control, two types of transitions are relevant:

- Transitions due to external actions

- State transitions due to control actions

The first type of transition results in "jumping" from one optimal goal path to the initial state of another optimal goal path while the second type of state transition does not result in deviations from one to another optimal goal paths. The control and collateral state sets of a state are intended to distinguish the states which are reachable from a certain state only via state transitions from the states which are reachable via transitions.

4.8.1 The control state set of a state

The *control state set* $K_+(S)$ relative to a state S is the set of states reachable from S by control actions alone. In other words, $K_+(S)$ consists of states belonging to the same optimal goal path as S and reachable from S. This is so because in the case when the goal paths are constructed from the elements of a well-determined state set each state is in only one optimal goal path.

4.8.2 The external state set of a state

The *external (collateral) state set* $K_-(S)$ relative to a state S is the set of states reachable from S by external actions alone. According to *property 4:6* each state with an external action has as its consecutive state the initial state of some optimal goal path. Now suppose that there are three states in S such as $S_i^p = ((z_i,\ c_i^p),\ u_{i,j}^{p,q})$, $S_i^p = ((z_i,\ c_i^p),\ u_{i,k}^{ext})$, and $S_m^t = ((z_m,\ c_m^t),\ u_{m,k}^{t,p})$ where the last state is an initial state. Then this initial state is in the collateral state set of both $S_i^p = ((z_i,\ c_i^p),\ u_{i,j}^{p,q})$, and $S_i^p = ((z_i,\ c_i^p),\ u_{i,k}^{ext})$.

The control and collateral state sets of a state have the following properties.

Property 4:7. *Each state in $K_-(S)$ belongs to only one optimal goal path.*

This property follows directly from the fact that each state in $K_-(S)$ is also a state in the well-determined state set S. Since every state in S, initial or not, can be in only one optimal goal path, it follows that each state in $K_-(S)$ is the initial state of only one optimal goal path.

Property 4:8. *The intersection of $K_-(S)$ with $K_+(S)$ is the empty state, i.e.,*

$$K_-(S) \cap K_+(S) = \{\phi\}.$$

This property follows directly from the definition of $K_-(S)$ and $K_+(S)$.

Proposition 4:6. *Let S_i and S_j be two arbitrary states in the same goal path. Then:*

$$K_-(S_i) \cap K_-(S_j) = \{\phi\}$$

Proof.
Assume that the intersection $K_-(S_i) \cap K_-(S_j) = \{S_k\}$. That means that S_k is a state reachable by some different external actions from both S_i and S_j. Thus, the configuration formula of S_i and the configuration formula of S_j are identical to the configuration formula of S_k hence, S_i and S_j have the same configuration formula, but different proper plant formulas. Since both of these states belong to a well-determined state set the fact that they have the same configuration formula, but different proper plant formulas contradicts the property of configuration specificity. Thus, $K_-(S_i) \cap K_-(S_j) = \{\phi\}$.

QED

4.9 Summary

In this chapter we have defined states with explicitly represented control configurations. This definition extends the notion of a *well-defined state* from Chapter 3 by augmenting it with configuration formulas. Furthermore, we have restricted the state set from Chapter 3 by introducing the so called *well-determined state set* characterized by the properties of syntactical completeness, integrity, specificity of control action and specificity of control configuration. A consequence of these properties is the elimination of non-optimal goal paths and of control cycles. We also showed that transitions due to external actions, (which are part of states in a well-determined state set), always result in transitions to initial states of optimal goal paths. Finally, we introduced the notions of a *control state set* and a *collateral state set* of a state and showed that these two state sets are distinct for each two states in a well-determined state set.

In the following chapter, we shall investigate the behavior of the GSO on a well-determined state set in the case of de-synchronization due to a number of different causes (VOA, ill-represented states, external actions, etc.) and we show how these causes can be distinguished from each other.

REFERENCES

[1]   Ravn, A.P., Rischel, H., Hansen K.M. Specifying and Verifying Requirements of Real-Time Systems, IEEE Trans on Software Eng. January 1993.

5

VIOLATIONS OF ONTOLOGICAL ASSUMPTIONS

In this chapter we show that the different causes for de-synchronization can be distinguished from each other when the state set of the object PC is well-determined. The causes for a de-synchronization considered here involve unexpected external actions, violation of ontological assumptions, ill-represented states, and timing. We do not deal with a de-synchronization due to expected external actions since these actions are defined a-priori and thus, their effects, though leading to a de-synchronization, are known in advance.

First we define the notions of ontological assumptions and violation of ontological assumptions. Then we introduce a modified version of the GSO from Chapter 3 denoted as GSO_{wds}. The GSO_{wds} is in fact a simpler version of the original GSO since the state set of the object PC the GSO_{wds} uses is well-determined and thus, each state and partial state (i.e., precondition pairs) is in only one optimal goal path. This in turn implies that the GSO does not have to consider different candidate control actions when completing a partial state since there is exactly one control action a partial state can be completed with. Finally, we analyse the behavior of the GSO_{wds} in the case of a de-synchronization due to the different causes from above. We show that the partial state, when completed by the GSO_{wds}, corresponds to a different state in the collateral state set of the state at which de-synchronization took place.

5.1 Ontological Assumptions

In this section we define the notions of ontological assumptions, violation of ontological assumptions, and we introduce the goal seeking operation GSO_{wds}. The later is used with well-determined state sets and is a modified, simpler version of the GSO from Chapter 3.

Ontological assumptions, as described in Section 3.5.2, are unstated assumptions underlying implicitly the validity of plant formulas. The separation of a plant formula y_i into a configuration formula c_i^p and a proper plant formula z_i, allow us to distinguish which is the particular part of y_i whose validity is based on ontological assumptions, namely z_i. Furthermore, we have that whenever the ontological assumptions implicit in the construction of z_i are violated, this proper plant formula cannot materialize and instead some other proper plant formula materializes.

Let S_i^p and S_j^q be two consecutive states in a well-determined state set S and let they be given as:

$$S_i^p = ((z_i, c_i^p), u_{i,j}^{p,q}) \in S$$

$$S_j^q = ((z_j, c_j^q), u_{j,k}^{q,r}) \in S$$

Thus, the two states above define a state transition $S_i^p \rightarrow S_j^q$.

The *ontological assumptions* relative to the state S_j^q are all the unstated assumptions which ensure that the following statement is true.

For any t there exists an $n \geq 1$ such that

- **IF** $S_i^p(t)=1$ (that is, $z_i(t)=1$, $c_i^p(t)=1$ and $u_{i,j}^{p,q}(t)=1$) **and**

- No other control action materialize in the time interval $[t,\ t+n]$ except $u_{i,j}^{p,q}$ **and**

- No expected and no unexpected external action occurs in $[t+1,\ t+n]$ **and**

- S_i^p and S_j^q have no ill-represented proper plant formulas and are not materialized due to a timing de-synchronization

- **THEN** $z_j(t+n) = 1$

The control considerations for the above definition are the following. When the control action $u_{i,j}{}^{p,q}$ is executed at t, it is certain that the configuration changes from $c_i{}^p$ to $c_j{}^q$. The object PC and its actuators have pre-defined, fixed properties and the purpose of ontological control is not the study of the interaction between the object PC and its actuators. As already described in Chapter 2, an actuator may have its own control system. The control at the level of the actuator realizes the intended output of the actuator, this intended actuator output being a goal state for the control system of the actuator. However, at the control level of the object PC to which the control actions affect the actuator, the intended actuator output is always achieved following the execution of a control action.

The situation is different for the intended plant output which the PC realizes due to the interaction between its actuators and the plant. Although the configuration formula $c_j{}^q$ materializes at $t+n$, it is only a conjecture that the proper plant formula z_j materializes at $t+n$ as well. This conjecture is made under a number of assumptions about the environment in which the interaction between the actuator output and the plant output takes place. Therefore, whenever these assumptions are violated, the proper plant formula z_i may not materialize at $t+n$. When the proper plant formula does not materialize due an interaction that contradicts the conjecture above, then a violation of ontological assumptions takes place.

A *violation of ontological assumptions* relative to the state $S_i{}^p$ is an unstated assumption which ensures that the following statement is true.

For any t and n

- **IF** $S_i{}^P(t)=1$ (that is, $z_i(t)=1$, $c_i{}^P(t)=1$ and $u_{i,j}{}^{p,q}(t)=1$) **and**

- No other control action materialize in the time interval $[t, t+n-1]$ except $u_{i,j}{}^{p,q}$ **and**

- No expected or unexpected external action occurs during $[t+1, t+n]$ **and**

- $S_i{}^p$ and $S_j{}^q$ have no ill-represented formulas and are not materialized due to a timing de-synchronization

- **THEN** $z_j(t+n) = 0$

5.2 The GSO$_{\textbf{wds}}$

The motivation for introducing a new GSO is the following. Recall that the GSO described in Section 3.4.3 performs three operations:

- The *state completion operation.* The GSO performs the state completion operation when the post-condition of a control action executed a t materializes at $t+n$ $(n{\geq}1)$. Using the notation from Section 3.4.3, this happens when the goal state is reached or either S^{∞} or $S^{\in}$ have a single element. The GSO completes then the materialized post-condition with the control action of the state in S^{∞}or $S^{\in}$. This operation is performed during steps 0, 2.2 and 3.2, described in Section 3.4.3.

- The *proper goal seeking operation.* The GSO performs this operation when it completes a partial state with a control action chosen among the control actions of several candidate states. This happens when either S^{∞} or $S^{\in}$ have more than one element. Each of the states in S^{∞} or $S^{\in}$ are on a goal path (optimal or non-optimal) towards the currently pursued goal state. The GSO uses for this operation the state index and the priority order on goal states as described in Section 3.4.3 steps 2.3 and 3.3.

- The *synchronization operation.* The GSO performs the synchronization operation during the synchronization step 3.1 (see Section $3.4.3$) when the currently pursued goal state is not reachable any more ($S^{\in}$ is empty). During this step, the GSO determines a new goal path using the priority relation on goal states, as defined in Section 3.4.2 and the plant formula materialized at t.

For a state set with explicitly represented control configurations, but which is **not** well-determined, the GSO described in Section 3.4.3 can still be used with the following modifications in the state notation: (i) a precondition y_i is modified to a precondition pair (z_i, c_t^p), and (ii) a control action $u_{i,j}$ is modified to $u_{i,j}^{p,q}$. However, these modifications in just notation preserve all the results obtained in Chapter 3 since at each t we have that:

$$y_i(t){=}1 \text{ iff } (z_i(t){=}1 \text{ and } c_t^p(t){=}1), \text{ (see Section 4.1.3) and}$$

$$u_{i,j}(t){=}1 \text{ iff } u_{i,j}^{p,q}(t){=}1 \text{ (see Section 4.2.1)}$$

i.e., the precondition pair (z_i, c_t^p) and the precondition y_i are logically equivalent and thus, the object PC materializes a control action with an explicitly represented control configuration, $u_{i,j}^{p,q}$ under the same conditions as for a control action without an explicitly represented configuration $u_{i,j}$.

However, a GSO acts differently on a well-determined state set with explicitly represented control configurations. Both the proper goal seeking operation and the synchronization step are different.

5.2.1 The proper goal seeking operation

The property of specificity of control action (Section 4.5.3) implies that a precondition pair (z_i, c_i^p) belongs to only one state (i.e. if $S_i^p=((z_i, c_i^p), u_{i,j}^{p,v})$ is a state of a well-determined state set S, then there is no other state $S_i^p=((z_i, c_i^p), u_{i,k}^{p,w})$ in S). Therefore the set $S^\in$ is always either empty or it has only one element. Moreover the property 4:5 requires that there are no non-optimal goal paths. Therefore the set S^∞ is always empty. These two properties imply that:

- When the state $S^\in$ has only one element, then the proper goal seeking operation is not required since there is exactly one candidate control action and thus, the state completion operation alone can complete a partial state with a control action.

- When both $S^\in$ and S^∞ are empty, then the proper goal seeking operation is not required since it is the synchronization operation which is activated in this case.

These two cases from above show that the GSO defined in Section *3.4.3*, when acting on a well-determined state set does not require the proper goal seeking operation. This is so simply because each state has only one consecutive state and thus, there is only one candidate control action which can be used to complete a given partial state.

5.2.2 The synchronization step

Recall that the GSO as defined in Section 3.4.3, performs the synchronization operation at step *3.1* when the currently materialized plant formula y_i cannot be completed with a control action $u_{i,j}$ so that the state $(y_i, u_{i,j})$ is on the currently pursued goal path (i.e., both $S^\in$ and S^∞ are empty). This means that the synchronization step has to find a new state on a different goal path. The GSO then uses the currently materialized plant formula y_i and the priority order on goal states.

For a well-determined state set S, the property of specificity of control action in 3.5.3 ensures that there is always a single state that has a certain plant formula y_i. Thus, there is always one and only one control action which can be used to complete the partial state $(y_i, -)$.

These properties of a well-determined state set allow us to define a GSO for well-determined state set called GSO_{wds}, as follows:

Let us assume that at time t, the state $S_i{}^p=((z_i, c_i{}^p), u_{i,j}{}^{p,q})$, is materialized, i.e. $z_i(t)=1$, $c_i{}^p(t)=1$ and $u_{i,j}{}^{p,q}(t)=1$. Let us assume moreover that at $t+n$ the precondition pair $(z_k, c_k{}^q)$ materializes (note that the configuration has the index q as required after the execution of the control action $u_{i,j}{}^{p,q}$). Then the GSO_{wds} proceeds through the following steps:

STEP 0

Step 0 is the state completion operation. The GSO_{wds} checks whether $j \equiv k$, i.e. if the precondition pair materialized at $t+n$ is the expected postcondition of the action $u_{i,j}{}^{p,q}$: $(z_j, c_j{}^q)$. According to the property of specificity of control actions (Section 4.5.3), there exists always only one state that has the expected post-condition pair $(z_j, c_j{}^q)$.

If $j \equiv k$ then the GSO_{wds} completes $(z_j, c_j{}^q)$ with the control action $u_{i,k}{}^{q,r}$ of the consecutive state to S_i which is $S_j=((z_j, c_j{}^q), u_{j,k}{}^{q,r})$. Thus, the control action $u_{j,k}{}^{q,r}$ is executed at $t+n$ and that result in a new partial state being materialized after an appropriate time interval. Then again, the GSO_{wds} continues from step 0 with this new precondition pair. If the precondition pair $(z_j, c_j{}^q)$ belongs to a goal state then step 0 completes the control action $u_{j,j}{}^{q,q}$ instead.

If $j \neq k$, the GSO_{wds} continues with step 1.

STEP 1

Step 1 performs a synchronization operation. The GSO_{wds} determines the goal path to pursue at $t+n$ using for this purpose the currently materialized precondition pair $(z_j, c_j{}^q)$.

At this step, the GSO_{wds} checks which state in S has the plant formula pair $(z_k, c_k{}^q)$. Let this state be $S_k=((z_k, c_k{}^q), u_{k,l}{}^{q,r})$. The property of specificity of control action (4.5.3) ensures that there exists exactly one state that has this precondition pair. The

GSO_{wds} completes $(z_k, c_k{}^q)$ with the control action $u_{k,l}{}^{q,r}$ and this control action is executed at $t+n$, i.e., $u_{k,l}{}^{q,r}(t+n)=1$. Then the GSO_{wds} continues from step 0 with the plant formula pair materialized after an appropriate time interval.

The GSO_{wds} has the following properties:

- If the GSO_{wds} starts from a state on the current optimal goal path to the current goal state G_k and the expected post-condition pair of each control action is always materialized, then the GSO_{wds} will achieve G_k via the current goal path. This property follows immediately from the fact that in the case of a well-determined state set there are no non-optimal goal paths (property 4:5). In this case, the GSO_{wds} will complete states with control actions by executing consecutively *STEP 0* for each state on an optimal goal path.

- If the state $S_i{}^p$ materializes at t and at $t+n$ the precondition pair $(z_k, c_k{}^p)$ materializes which is not the expected post-condition pair of the control action of $S_i{}^p$, then the GSO_{wds} will complete $(z_k, c_k{}^p)$ with a control action so that the completed state is a state in $K_-(S_i{}^p)$. This follows immediately from the property 4:6, the definition in 4.7.2 of a collateral state and the definition of the synchronization step of the GSO_{wds}.

The situation described above occurs in the case of a de-synchronization, i.e., the materialized partial state (the precondition pair) is not the expected precondition pair of the consecutive state of $S_i{}^p$. For well-determined state sets, as shown above, a de-synchronization always results in a completed state that belongs to the collateral state of a given state. However, there exist different causes for a de-synchronisation and our purpose is to find a criteria to distinguish between them. In the following section we show that this distinction is given by a relation between the state that is the argument of the collateral state set and the expected state that did not materialize.

5.3 Distinguishing Between De-synchronization causes

When a de-synchronization takes place due to one of the causes described in Section 3.5.2, the proper plant formula of the expected postcondition pair does not materialize, but the proper plant formula of some non-consecutive state materializes instead. This raises the following questions:

- Which are the states that can materialize after a de-synchronization? In other words, can we define a subset of S such that each state that materializes following a de-synchronization belongs to this subset?

- Is this subset characteristic for each type of de-synchronization? In other words, are there subsets of it corresponding to each cause for a de-synchronization such that these subsets are non-empty and disjoint ?

both these questions have a positive answer, then the different causes for a de-synchronization can be identified with these particular subsets.

To answer the questions from above, we analyse in the next section each type of de-synchronization from Section 3.5.2.

5.3.1 Ontological De-Synchronization

We seek here a subset of S with states that can materialize after violations of ontological assumptions.

Let us assume that at time t the current materialized state is S_t^p and the consecutive state to S_t^p is S_j^q. Both states are on an optimal goal path to say, S_g. Due to violations of ontological assumptions of S_j^q, at time $t+n$, the precondition pair of S_j^q does not materialize, but the GSO_{wds} materializes the state S_l^r instead (Figure 22). Using the following notation for the states:

$$S_t^p = ((z_i,\ c_i^p),\ u_{i,j}^{p,q})$$

$$S_j^q = ((z_j,\ c_j^q),\ u_{j,k}^{q,s})$$

$$S_l^r = ((z_l,\ c_l^r),\ u_{l,m}^{r,t})$$

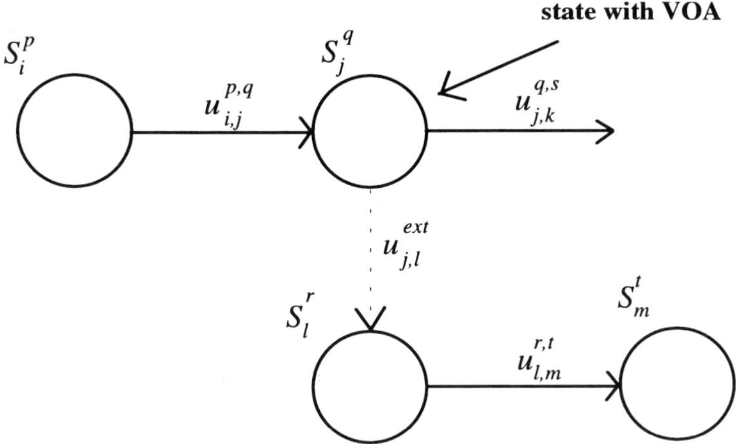

Figure 22 Ontological de-synchronization.

we have the following proposition:

Proposition 5:1 *Let S_i^p and S_j^q be two states of a well-determined state set S defining a state transition $S_i^p \rightarrow S_j^q$. If S_i^p is materialized at t and S_j^q has a violation of ontological assumptions at t+n, then the GSO materializes at t+n a state S_l^r such that $S_l^r \in K_{-}(S_j^q)$.*

Proof.

According to the definition of a control action, $u_{i,j}^{p,q}$ realizes always the control configuration c_j^q, i.e., $c_j^q(t+n)=1$. This means that the state S_l^r materialized instead of the state S_j^q has a configuration formula with a upper index $r\equiv q$, i.e., $S_l^r = ((z_l, c_l^q), u_{l,m}^{q,t})$. However, according to the definition of a violation of ontological assumptions, the proper plant formula of S_j^q does not materialize i.e. $z_j(t+n)=0$, but some other proper plant formula materializes. Let this plant formula be $z_l \neq z_j$ and $z_l(t+n)=1$. Let us examine now which are the states in S, besides the state S_j^q that can have the configuration c^q. The states in S can be either states that are initial to optimal goal paths, or states that are not initial to optimal goal paths. According to the property of specificity of control configuration the only other state in S with the same configuration c^q as S_j^q but a different proper plant formula is a state initial to an optimal goal path. According to the definition of collateral state set of a state, the states that have the configuration formula c^q and are initial states to an optimal goal path are the states in $K_-(S_j^q)$.

Q.E.D.

Definition 5:1 The de-synchronization that materializes a state in the collateral state set $K_-(S_j^q)$ after the state S_i^p is materialized and $S_i^p \rightarrow S_j^q$, is called an *ontological de-synchronization*.

5.3.2 Unexpected External Action De-Synchronization

Let us assume again a well-determined state set S and two states $S_i^p=((z_i, c_i^p), u_{i,j}^{p,q})$ and $S_j^q=((z_j, c_j^q), u_{j,k}^{q,r})$ in S such that $S_i^p \rightarrow S_j^q$. Moreover, let the state S_i^p be materialized at t, $S_i^p(t)=1$, and let us assume that the configuration change from c_i^p to c_j^q occurs after a number of n inner state transitions of the state S_i^p. Therefore, for each t in $[t, t+n-1]$ we have that $S_j^q(t)=0$. Let us assume that an unexpected external action occurs at t'. where t' is in $[t, t+n-1]$. As already described in Chapter 4 this unexpected external action has the following effect:

- Due to the occurrence of the unexpected external action, a proper plant formula z_l becomes true at $t+n$.

- The state S_i^p does not materialize at $t+n$ i.e., $z_i(t+n)=0$.

- The configuration formula in the precondition pair of the state S_i^p remains unchanged at $t+n$.

Therefore, according to the property of specificity of control configuration, the states that have the same configuration c^p as S_i^p but a different plant formula than z_i are only those initial to goal paths. Using the definition of the collateral state set of a state, these states are those in $K_-(S_i^p)$.

In conclusion, the states that GSO_{wds} can materialize after the occurrence of an unexpected external action, while the state S_i^p, is in inner state transitions are the states in $K_-(S_i^p)$.

Definition 5:2 The de-synchronization that materializes a state in the collateral state set $K_-(S_i^p)$ after the state S_i^p is materialized and $S_i^p \rightarrow S_j^q$, is called an external *de-synchronization*.

We should mention here that the external de-synchronization due to an expected external action also results in the materialization of a state in $K_-(S_i^p)$. This is so since any external action keeps the configuration formula c^p unchanged and just results in a change of the expected proper plant formula. However, in this case one knows exactly which this proper plant formula is since the effect of an expected external action on the plant is specified a priori. Thus, one knows that the materialization of the plant formula z_k instead of the expected plant formula z_j is due to an expected external action $u_{i,k}ext$.

The following proposition shows that the states that can materialize after a de-synchronization due to VOA are distinct from the states that can materialize after an external action (expected or unexpected).

Let $S_i^p=((z_i,\ c_i^p),\ u_{i,j}^{p,q})$ and $S_j^q=((z_j,\ c_j^q),\ u_{j,k}^{q,r})$ be two state in S such that $S_i^p \rightarrow S_j^q$. We assume now that the state S_j^q has a violation of ontological assumptions. Let the state S_i^p be materialized at t. We assume that the configuration change from c_i^p to c_j^q occurs after several inner state transitions of the state S_i^p, for instance let the configuration c_j^q be materialized at $t' > t$. However, due to VOA, the proper plant formula of the state S_j^q does not materialize.

Proposition 5:2 *For a well-determined state set S, the set of states GSO_{WDS} can materialize after an expected or unexpected external action is disjoint with the set of states it can materialize after a VOA.*

Proof.
The states that can materialize following the occurrence of an expected or unexpected external action are those in $K_-(S_i^p)$ The states that can materialize after a VOA are those in $K_-(S_j^q)$. Then property *4:6* states that:

$$K_-(S_i^p) \cap K_-(S_i^q) = \{\varnothing\}.$$

Q.E.D.

5.3.3 Ill-Represented Formula De-Synchronization

A de-synchronization due to an ill-represented proper plant formula of a state $S_k^p=((z_k, c_k^p), u_{k,m}^{p,t})$ occurs when at time t the control action $u_{k,m}^{p,t}$ is executed, but the proper plant formula in (z_k, c_k^p) of S_k^p does not correspond to the actual plant output at t. The precondition (z_k, c_k^p) is interpreted to true with outdated data and thus, the actual plant output at t is actually represented by some other proper plant formula different from z_k.

Let us consider the states S_i, S_j, S_k and S_m in the well-determined state set S such that $S_i^p=((z_i, c_i^p), u_{i,j}^{p,q})$, $S_j^q=((z_j, c_j^q), u_{j,k}^{q,r})$, $S_k^p=((z_k, c_k^p), u_{k,m}^{p,t})$, $S_o^t=((z_o, c_o^t), u_{o,p}^{t,x})$, the state transitions $S_i^p \to S_j^q$, $S_k^p \to S_m^t$,and the transitions S_i^p --> S_k^p and S_m^t --> S_o^t. (see **Figure 23**). Moreover we consider the simpler version of the goal seeking operation, GSO_{wds}.

The state materialized at t is S_k^p and let us assume that S_k^p is ill-represented. That is, the state S_k^p has a proper plant formula z_k that does not correspond to the current plant output though it is interpreted to true. However, there is another state, S_i^p, which has the same configuration as S_k^p and its proper plant formula is actually true for the current plant output at t. In other words, the precondition pair of S_k^p materializes at t although the precondition pair that should materialize at t is that of S_i^p.

An ill-represented state may seem to cause an ontological de-synchronization due to the following reasons. The precondition pair (z_k, c_k^p) materializes at t. However, the actual plant output is represented by (z_i, c_i^p). Since the control action $u_{k,m}^{p,t}$ is executed at t, but the proper plant formula z_k its precondition pair (z_k, c_k^p) does not correspond to the actual plant output at t, the plant formula z_m in its post condition pair (z_m, c_m^t) cannot materialize at some time $t' > t$.

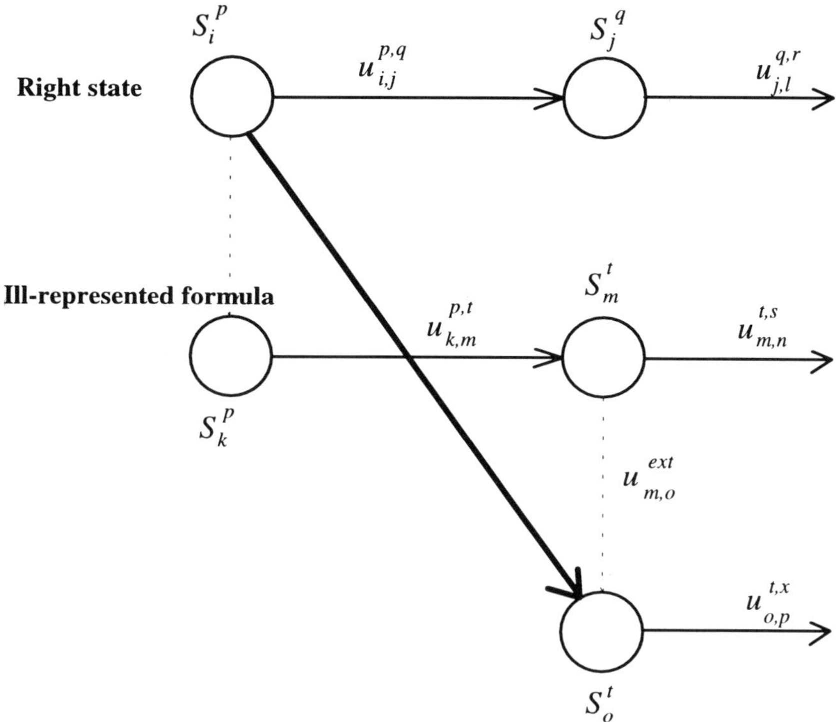

Figure 23 Ill-represented formula de-synchronization.

Since the control action changes the configuration from c^p to c^t, the configuration $c_m{}^t$ is realized indeed, but the proper plant formula z_m is not materialized. Therefore, the state that can materialize in this case should be a state in $K_-(S_m{}^t)$.

However, the following proposition shows that this is not the case.

Proposition 5:3. *For a GSO$_{wds}$ acting on a well-determined state set, the state materialized after a de-synchronization due to an ill-represented state $S_k{}^p$ is different from the state materialized by an ontological de-synchronization on $S_k{}^p$.*

Proof.

Recall that an ontological de-synchronization on $S_k{}^p$, with $S_k{}^p \rightarrow S_m{}^t$, is identified with the materialization of a state in $K_-(S_m{}^t)$ which state has the precondition pair $(z_l, c_k{}^t)$.

Let us assume that the precondition pair materialized at t is (z_k, c_k^p). The GSO$_\text{wds}$ completes this precondition pair at step 0 with the control action $u_{k,m}^{p,t}$ of the state S_k^p and executes this control action at t, i.e., $u_{k,m}^{p,t}(t)=1$. However, since the actual plant output corresponds to the precondition pair (z_i, c_i^p), we have that the actual state that is materialized at t is $((z_i, c_i^p), u_{k,m}^{p,t})$. This state is not a well-defined state since the index i of the plant formula in its precondition pair (z_i, c_i^p) does not correspond to the first lower index k of the control action $u_{k,m}^{p,t}$. The index k means that the control action $u_{k,m}^{p,t}$ must have in its precondition pair the proper plant formula z_k rather than z_i in order to realize the intended plant output z_m. Otherwise the latter proper plant formula cannot be materialized after the execution of the control action $u_{k,m}^{p,t}$. Since a well-determined state set consists only of well-defined states we have that the state $((z_i, c_i^p), u_{k,m}^{p,t})$ does not belong to the well-determined state set to which the states S_k^p and S_m^t belong. As shown in Chapter 4, the consecutive state of a state not in a well-determined state set is not in this well-determined state set either. Thus, the state that can be materialized after a de-synchronization due to an ill-represented proper plant formula of the state S_k^p is not in the well-determined state set. At the same time an ontological de-synchronization results in the materialization of a state (initial to a an optimal goal path) in the well-determined state set. Hence an ill-represented formula de-synchronization and an ontological de-synchronization result in the materialization of different states.
Q.E.D.

To summarize, a de-synchronization due to an ill-represented state results in the materialization of a state not in S and thus, this state is not on any optimal goal path. In this manner we can distinguish this particular cause for a de-synchronization from ontological and external action de-synchronizations which all result in the materialization of initial states on optimal goal paths.

5.3.4 Timing De-Synchronization

A de-synchronization due to *timing* or *timing de-synchronization* occurs when the so called *state density property* is violated. The state density property states that if S_i^p is a transient state (see Section 4.4.2) and S_j^q is the consecutive state of S_i^p then there is **no** state that can be materialized in the time interval $[t, t+1]$.

We describe here this type of de-synchronization since it may give states in K_- (S_j^q) as in the case of an ontological de-synchronization. Assume that the state materialized at time t is S_t^p, and that the consecutive state to S_t^p is $S_j^q = ((z_j, c_j^q), u_{j,k})$. Assume moreover that there is a state S_l^q reachable from S_j^q via a transition due to an expected external action. A timing de-synchronization means that the state S_l^q materializes at time $t+1$ instead of the state S_j^q simply because it takes a longer time to materialize S_j^q. For the GSO_{wds}, this appears as a genuine de-synchronization due to an ontological de-synchronization (since S_t^p and S_l^q have the same configuration formula c^q but different proper plant formulas), although it is in fact due to the wrong choice of the sampling interval. If it takes even a longer time to detect the materialization of S_j^q then a number of transitions and/state transition can be materialized before the time when this state is expected to materialize.

5.4 Post Synchronization Behavior

In this section we will briefly discuss the post de-synchronization behavior of the GSO_{wds} after a de-synchronization due to the above described causes.

In the case of external de-synchronization the state materialized is an initial state on some optimal goal path and the consequent interpretation and completion of states will continue on this optimal goal path. Thus, if no other de-synchronization occurs such that the initial state on the aborted goal path is materialized, the goal state of this optimal goal path can never be achieved. If too many external de-synchronizations occur, this lead to a situation in which the object PC will be jumping between the initial states of optimal goal paths without being able to complete them.

In the case of an ontological de-synchronization on a state say, S_m^r the following situation may occur. The plant formula z_m of this state is a Boolean function of a number of constrained plant signals. A violation of ontological assumptions on z_m implies that some of these constrained plant signals do not still reflect the actual plant outputs they were intended to represent. If these constrained plant signals are part of other plant formulas, these will also be affected by the violation of ontological assumptions on z_m. Thus, the violation of ontological assumptions on certain constrained plan signal can affect more than one state. In the worst case, when all states on the optimal goal paths have a violation of ontological assumptions the object PC will be just jumping from one initial state to another without being able to achieve any goal state at all. Since there are a finite number of initial states it may, after a time, run out of new initial states and thus, run into a cycle consisting only of initial states.

In the case of an ill-represented formula de-synchronization the materialized state is not on any optimal goal path. The consecutive state of a state with ill-represented formula is not on any optimal goal path either. Thus the object PC even if able to interpret and complete such states, will not be able to achieve none of its goal states.

In this context, it is of extreme importance that the object PC is made aware of the fact that there is a de-synchronization, what the exact cause of this de-synchronization is, and the possible ways for synchronization. In Chapter 7 we describe a control architecture that can exhibit a VOA.

5.4.1 Summary

In this chapter, we showed that the GSO_{wds} when acting on a well-determined state set can distinguish between a number of different causes for a de-synchronization by means of the materialization of collateral state sets specific for each cause. The results are summarised in the following table:

Table 3

De-synchronization type	De-synchronized state transition	Materialized state after de-synchronization
Ontological	$S_i^p \rightarrow S_j^q$	in $K_-(S_j^q)$
External action	$S_i^p \rightarrow S_j^q$	in $K_-(S_i^p)$
Timing	$S_i^p \rightarrow S_j^q$	in $K_-(S_j^q)$
Ill-represented formula	$S_i^p \rightarrow S_j^q$	not in S

5.5 Example

The following example (**Figure 24**) illustrates the behavior of an object PC under VOA. A radar tracking system has two positioning mechanisms. First, a rough set-up position of the target object is given to the radar and a regulator R1 brings the radar into this position. For this purpose, the regulator measures the position using a mechanical device. After the rough position is reached, another regulator, R2, locks the radar to the target and a relative position error input is used which does not require any more the mechanical positioning measurement. The ontological assumption is that when the regulator R1 has reached the set-up position, then the object is in the locking range of the regulator R2. However, a violation of this assumption occurs: the mechanical measuring device is not fixed to the radar axle and has an offset error, i.e., instead of the real angle α it gives an angle $\alpha+\varepsilon$.

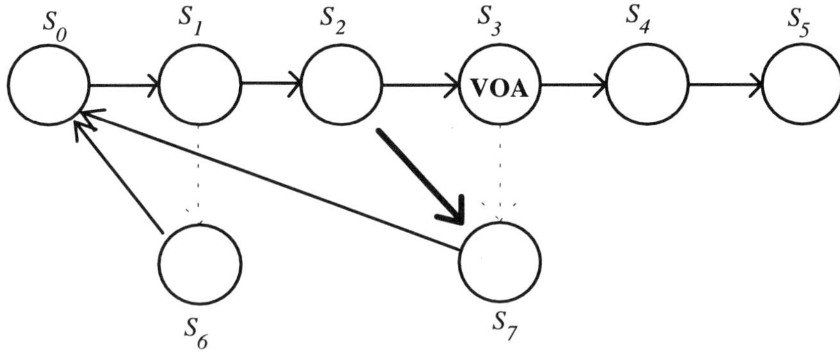

Figure 24 State set of a radar tracking control system (ii).

The states are the following:

S_0
z_0 = (reference value received)
c_0 = (system stopped)
$u_{0,1}$= (start regulator R1)

S_1
z_1 = (in range for R1) $\wedge\neg$ (in range for R2)
c_1 = (R1 is ON) $\wedge$ (R1 is active)
$u_{1,2}$= (regulator action R1)

S_2
z_2 = (in range for R2)
c_2 = (R1 is ON) $\wedge$ (R1 not active)
$u_{2,3}$= (stop R1) $\wedge$ (start R2)

S_3
z_3 = (in range for R2) $\wedge \neg$ (range for goal)
c_3 = (R2 is ON) $\wedge$ (R2 is active)
$u_{3,4}$= (regulator action R2)

S_4
z_4 = (in range for goal)
c_4 = (R2 is ON) $\wedge$ (R2 not active)

$$u_{45} = \text{(stop R2)} \wedge \text{(start goal)}$$

S_5

$$z_5 = \text{(goal fulfilled)}$$
$$c_5 = \text{(goal configuration is ON)}$$
$$u_{5,5} = \text{(wait in inner loop)}$$

S_6

$$z_6 = \text{(outside position for R1)}$$
$$c_6 = \text{(R1 is ON)} \wedge \text{(R1 is active)}$$
$$u_{6,0} = \text{(reset action)}$$

S_7

$$z_7 = \text{(outside working position for R2)}$$
$$c_7 = \text{(R2 is ON)} \wedge \text{(R2 is active)}$$
$$u_{7,0} = \text{(stop R2)}$$

The sequence of state transitions is the following:

1. The system starts at S_0 with a reference value for the regulator R1.

2. During state S_1 the regulator R1 brings the radar into the reference position. When the reference position is reached, the regulator has no more control action, so the configuration c_1 is not true any more.

3. The state S_2 stops R1 and starts R2.

4. The ontological assumption is that the state S_3 can always be materialized since the radar is now in range and R2 can be started.

5. This assumption is not true due to the measurement device offset, which gives a large error position error, identical to one obtained after an external disturbance. The state S_7 was intended as a collateral to S_3 if some large noise makes the tracking with R2 impossible. Due to the ontological error, the an object PC desynchronizes into the state S_7 yet no external action has occurred.

6. Since the error is large, S_7 stops the regulator R2 and assumes that S_1 can be materialized. However the proper plant formula of S_1 cannot materialize since z_1 has the condition that the radar is not in range to start the regulator R2, but that is not true, the radar is in range for R2 since the plant formula z_2 is still true.

7.   The regulator materializes S_2 and the cycle continues at S_7.

The regulator stays indefinitely in this loop.

6

DETECTING VOA ON A NON WELL-DETERMINED STATE SET

6.1 Introduction

In this chapter we seek a solution to the following problem. The results from Chapter 3 show that violations of ontological assumptions do not lead to a recognizable behavior for non-well determined state sets. However, in practical control applications, it is often difficult to design object controllers with well-determined state sets. The next Section, 6.2, presents some of the circumstances that contribute to these difficulties.

A solution to this problem is to transform the state set of the object controller into a new state set that is well determined. Then, violations of ontological assumptions can be detected on this newly created state set using the results from Chapter 5.

The controller with the new state set is required to act identically as the object controller using the initial state set in cases when no problematic control situations occur. Moreover, in cases of problematic control situations, the controller with the new state set should have the specific behavior described in Chapter 5.

The main result in this chapter is the proof that, under certain constraints, the state set S of the object controller together with the goal seeking operation defined in Section 3.4.1 can aggregate into a state set that is well determined. This state set is called S_{gso}. Since S_{gso} is well determined, a controller with this new state set can use the goal seeking operation GSO_{wds} defined in (Section 5.1). We show that a controller with the pair (S, GSO) performs identically as a controller with the pair

(S_{gso}, GSO_{wds}) in cases when no de-synchronizations occur or there are only de-synchronizations due to expected external actions. Moreover, in cases of violations of ontological assumptions or unexpected external actions, a controller with the pair (S_{gso}, GSO_{wds}) has distinct behavior for each one of the above types of de-synchronization. By contrast, a controller with the pair (S, GSO) cannot show this distinction. In the last part of Chapter 6 we discuss a number of implications of the state transformation described above.

We remark that the transformation of the state set does not mean for instance a new control layer on the top of the initial object controller. After the transformation of the state set, the controller simply acts with the new state set.

Although both state sets are based on the same control knowledge, the new state set has a different abstraction level than the initial one. Therefore, the meaning of what is a "plant formula", "control action", "configuration", "state transition" changes with the state set transformation.

6.2 Why State Sets Are Not Well Determined?

Real world control applications most often do not have well-determined state sets due to the following inherent limitations:

- For control applications of non-trivial complexity it is a complex task to delimit precisely the controller configuration from the plant formulas. For example one may consider that all the process values the controller reads are *plant outputs* (i.e. all actuators may be considered as part of the plant) while from another perspective, all the plant may be considered as an actuator.

- Normally an object controller acting in a hierarchical control system uses the current state of the peer controllers acting on the same plant in its precondition formulas. Due to its size, the state set of the peer controllers cannot be incorporated within the state set of the object controller. Therefore, when a de-synchronization occurs, the object controller cannot perform a goal seeking or a synchronization operation based on the current state of each peer controller. Instead, the GSO applies some domain-dependent reasoning on the state set of the external control system that reduces the problem size. Most often the reduction is a result of a trade-off between price, complexity and the risks implied by a possibly wrong synchronization. Therefore the state set S of the object controller is by purpose not well-determined and the control relies on the existence of a GSO to complete the state in cases of de-synchronizations.

- The *state density property* (see Section 5.3.4) cannot be guaranteed due to the following two reasons: (1) A real world application requires to manage changes between the so called control modes: *manual mode* and *automatic mode.* The control mode can change at any time and inherently the controller in manual mode drops some intermediary states. (2) In a hierarchical control system, different parts of the system work in parallel. For a large state set, the number of combinations of state transitions that may occur simultaneously is very large such that normally the state density property cannot be guaranteed. It is not in the scope of this book, but it can be shown (author's unpublished paper) that in general, after a de-synchronization due to failures of the state density property, a synchronization cannot be done when the state set of the object controller is well-determined. Therefore, as in the previous case, a GSO performs a synchronization based on a domain-dependent reasoning that reduces the problem size.

6.3 Effective Control Paths

The state transformation we seek is based on the concept of effective control paths. In this section we define what is an effective control path and we show some properties for effective control paths that are required later on for the definition of the state set S_{gso}.

6.3.1 The Definition Of An Effective Control Path

In this section we define an *effective control path* as the ordered sequence of successive control actions and expected external actions that can effectively occur in a control application. It is apparent from what was described in Chapter 3 that all the possible sequences of control and expected external actions are implicit in the state set S and in the algorithm of the GSO. The ordered sequence is defined under the assumption that no de-synchronizations occur except those due to expected external actions.

The state set S and the GSO we use in this section are those defined in Chapter 3. Thus the state set S is well-defined but not necessarily well-determined. Moreover S is not required to have explicitly represented configurations. Therefore we use the notation $(y_i, u_{i,j})$ for a state instead of the notation with configurations $((z_i, c_i^p), u_{i,j}^{p,q})$.

The set of goal states of S is $G=\{G_1, G_2,...,G_g\}$. The GSO has a priority order among the goal states $G_1 > G_2 > ...> G_g$, as defined in the extended image state set of the GSO (see 2.4.2).

We distinguish the following particular states.

The state $S_1 = (y_1, u_{1,2})$ is the first state of the optimal goal path leading to G_1, where G_1 is the goal state with the highest '>' priority. This initial state has the property that it is not consecutive to any other state in S. In other words, the plant output y_1 of S_1 corresponds to an initial plant state that exists during the *cold-start-up* defined in Chapter 2. We assume that a plant has a unique initial state: if there would be several initial states, each being in a goal path towards G_1, then G_1 is reached by several optimal goal paths. However, we assume that there is a unique optimal goal path towards a goal state.

According to the properties of goal states in Section 3.2.2, a goal state does not have a consecutive image state. However, some goal states may have consecutive states due to expected external actions, while some other goal states may not have such consecutive states.

In what follows, we define the set of all the possible sequences of control and expected external actions that may occur for a controller with a given pair (S, GSO).

Definition 6:1 *Effective control path.* An *effective control path,* is a sequence of literal symbols $u_{i,j}^{\circ}$ separated by the '$\Rightarrow$' symbol such that:

- Each symbol $u_{i,j}^{\circ}$ denotes either a control action $u_{i,j} \in S$ or an expected external action $u_{i,j}^{ext}$.

- The first symbol in the sequence is some $u_{1,i}^{\circ}$ $i=1,2,...,n$ (that is, $u_{1,i}^{\circ}$ corresponds to either $u_{1,i} \in S$ or a known $u_{1,i}^{ext}$). The first index of the first action symbol is always 1 while the other index can be any integer i such that $u_{1,i}$ exists in S or $u_{1,i}^{ext}$ is a known expected external action.

- Two consecutive action symbols in the sequence are separated by the '$\Rightarrow$' symbol, as in $u_{i,j}^{\circ} \Rightarrow u_{j,k}^{\circ}$. The indices of the two consecutive action symbols are such that j, the second index of the left action symbol is identical to the first index of right action symbol.

- The last literal in the sequence is an action symbol $u_{g,g}^{\circ}$ that denotes the control action $u_{g,g}$ of a goal state $S_g = (y_g, u_{g,g})$. This goal state has one of the following properties:

    - $u_{g,g}$ is such that there is no known expected external action $u_{g,k}^{ext}$ or,

- $u_{g,g}$ is such that there exists an expected external action $u_{g,l}{}^{ext}$ but $u_{g,l}{}^{ext}$ is already an action symbol in the current effective control path prior to $u_{g,g}{}^{\circ}$.

- Each action symbol appears only once.

(Note that $u_{g,g}$ is <u>defined</u> when there exists an $u_{g,l}{}^{ext}$ which is already in the current effective control path, but $u_{g,l}{}^{ext}$ is <u>not</u> represented twice).

Notation: An effective control path is denoted as T_m, $m=1, 2,..., e$.

Remark 1. The condition that two consecutive action symbols are $u_{i,j}{}^{\circ} \Rightarrow u_{j,k}{}^{\circ}$, means that at two time instances t and $t' > t$, states materialize according to one of the situations below:

$$(y_i, u_{i,j})(t)=1 \text{ and } (y_j, u_{j,k})(t')=1 \text{ or}$$

$$(y_i, u_{i,j}{}^{ext})(t)=1 \text{ and } (y_j, u_{j,k})(t')=1 \text{ or}$$

$$(y_i, u_{i,j})(t)=1 \text{ and } (y_j, u_{j,k}{}^{ext})(t')=1 \text{ or}$$

$$(y_i, u_{i,j}{}^{ext})(t)=1 \text{ and } (y_j, u_{j,k}{}^{ext})(t')=1$$

Other cases, such as actions that materialize after de-synchronizations which are not due to expected external actions, are excluded by the definition of the effective control path. (Such actions do not have the required property for their indices).

Remark 2. The fact that action symbol appears only once in a sequence implies that not all the control actions and expected external actions are represented in effective control paths: the control actions and expected external actions that may lead to cycles are not represented. For example let us assume that the following two goal paths have their control actions in the same effective control path:

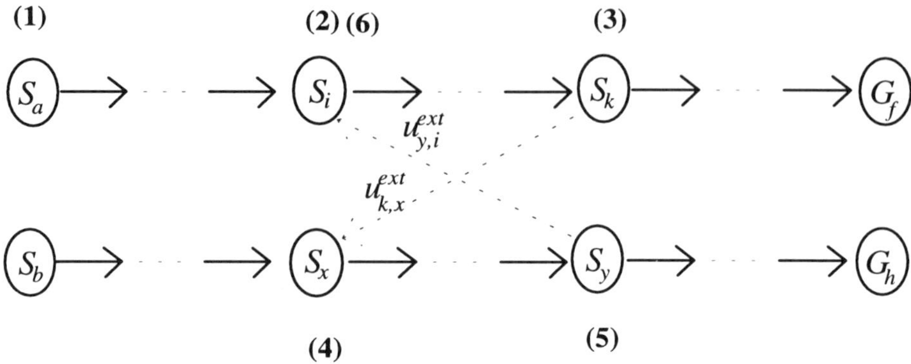

S_a, S_b being initial states and G_f, G_h being goal states. Let us assume that there are two expected external actions $u_{y,i}^{ext}$ and $u_{k,x}^{ext}$. According to the last condition for effective control paths, if the expected external action $u_{k,x}^{ext}$ is in the current effective control path, then the expected external action $u_{y,i}^{ext}$ is not represented in the effective control path since then the control action of the state S_i would appear twice in the effective control path (see in the figure above the order in which states are add to the effective control path, shown by numbers enclosed in parentheses). The reason why a control or expected external action appears only once, is that we are interested only in the reachability of the goal states from the initial state, and thus the repetition of the same state or the same sequences of states is not relevant. If a state or a sequence of states is materialized once without ontological violations, then it is not relevant how many times this sequence is materialized again. Another case which leads to repetitions was already described in Section 3.4.4. There a goal path may have a finite number of repetitions of the same sequences of states. The GSO augments these states with additional knowledge that limits the number of repetitions. The control action belonging to states in the cycle are represented again only once in the effective control path.

Remark 3. We shall use the notation $u_{i,j}$ or $u_{i,j}^{ext}$ in effective control paths for $u_{i,j}^{\circ}$ whenever the type of the action (i.e. control action or expected external action) is relevant in the context.

The set of all the effective control paths corresponding to a pair (S, GSO) is denoted as **T** and it is:

$$\mathbf{T} = \{T_1, T_2, ..., T_e\}$$

We show now how to construct the set of effective control paths when S and GSO are known.

6.3.2 Constructing The Set Of Effective Control Paths

The construction of the set of effective control paths is done as follows. The goal seeking operation described in Section 3.4.3 is an algorithm that has as input a partial state $(y_i, \_)$ and as output a control action $u_{i,j}$ that belongs to a state $(y_i, u_{i,j})$ in S. To construct the effective control path, we execute the algorithm off-line starting with the initial state $(y_1, \_)$ and we consider all the possible combinations for how the expected external actions may materialize. However, since no de-synchronizations are implied in the definition 5.1 except those due to expected external actions, not all the steps in Section 3.4.3 are used for effective control paths. The construction requires only the state completion operation of the object controller and of the proper goal seeking operation (that is, the synchronization steps - see Section 5.2.1 - are not used). In this way, the control actions the GSO completes, together with each combination of expected external actions, define a unique effective control path.

A summary of the GSO steps from Section 3.4.3, that are relevant for the construction of effective control paths is shown below. The intermediary steps of the algorithm that construct the sets of consecutive states S', S^∞ and $S^\in$ are left aside.

Let the present goal state be G_k and the partial image state the GSO has to complete be $(y_i, -)$.

STEP 0: The current goal state is reached, $i \equiv k$. The GSO completes the partial state with $u_{k,k}$ such that the current goal state G_k is reached. Then the current goal state changes to G_{k+1}, which is the next one in the '>'-priority order.

STEP 2.2: There is a state $(y_i, u_{i,r})$ which is the **unique consecutive in the current optimal goal path**. The GSO completes the partial state $(y_i, -)$ with the control action $u_{i,r}$.

STEP 2.3: There are **several states in the current optimal goal** path towards the same current goal state G_k. The GSO selects the state with the lowest index and completes its control action in the partial state $(y_i, -)$.

STEP 3.2: There is a unique image state $(y_i, u_{i,r})$ which is on a **non-optimal goal path** towards G_k. The GSO completes the partial state $(y_i, -)$ with a control action

$u_{i,r}$ such that the state $(y_i, u_{i,r})$ is the unique consecutive state towards the same current goal state G_k, on a non-optimal goal path.

STEP 3.3: There are **several states on non-optimal goal paths** towards the same current goal state. The GSO completes the partial state $(y_i, \text{-})$ with a control action $u_{i,r}$ of the state $(y_i, u_{i,r})$, selected among these states using the priority of goal states in the OGS and the state indices.

The synchronization step 3.1 it is not used for the construction of effective control paths. Recall that the STEP 3.1 changes the current goal path G_k to G_{k+1} when there are no available consecutive states towards the goal state G_k on the optimal goal path, nor on non-optimal goal paths.

Using the GSO steps described above, we construct the set of effective control paths with the following procedure:

- The first effective control path, denoted T_1, contains the control actions that correspond to complete goal paths. The effective control path T_1 is constructed using the GSO as follows. The first input to the GSO is the partial state $(y_1, \_)$ with y_1 being the precondition formula of S_1. The steps STEP 2.2 or STEP 2.3 are applied successively starting with $(y_1, \_)$ and control actions are attached to the effective control path that belong to states in the optimal goal path towards G_1. When the control action of the goal state G_1 is in T_1, STEP 0 is applied and the current goal path changes to G_2 as follows. Let G_1 be $(y_k, u_{k,k})$. If there exists an expected external action $u_{k,i}^{ext}$ such that $(y_k, u_{k,i}^{ext})$ has a consecutive state S_i belonging to the goal path towards G_2, then the expected external action symbol $u_{k,i}^{ext}$ is attached to T_1. The steps STEP 2.2 or STEP 2.3 are applied again starting with y_i and the control action for states in the optimal goal path towards G_2 are attached to T_1. When the control action of the goal state G_2 is in T_1, STEP 0 is applied and the current goal state changes to G_3. The procedure above continues until the control actions of all the states in goal paths, up to a goal state G_m, are in T_1. The goal state $G_m = (y_m, u_{m,m})$ has the following property: there is no expected external action $u_{m,p}^{ext}$ or alternatively, the expected external action $u_{m,p}^{ext}$ is already in T_1.

- The construction of another effective control path starts as above with the partial state $(y_1, \_)$ as an input to the GSO. At some state, in one of the optimal goal paths, a completion is done with the control action of a non-optimal goal path instead of the control action of the optimal goal path. Let us assume that using the procedure above, the goal state of the current optimal goal path is G_n ($n <$ m). Starting with the first state of the optimal goal path towards G_n, a number of control actions are attached to the effective control path, using STEP 0 and STEP 2.2 or STEP 2.3 as above. Let us assume that the current action symbol attached to the effective control path is $u_{i,j}$ of the state $(y_i, u_{i,j})$. However, at this state, the expected external action $u_{i,k}{}^{ext}$ is attached to the effective control path and the input to the GSO is now the partial state $(y_k, \_)$. The STEP 3.2 or STEP 3.3 can be now applied such that the state $(y_k, u_{k,l})$ belongs to a non-optimal goal path towards G_n.

- The two procedures above are repeated for each possible combination of expected external actions, and thus we obtain every possible *effective control path* that can be constructed with S and the GSO. Each T_i is distinct and is obtained using one of the possible combinations of STEP 0, STEP 2.2, STEP 2.3, STEP 3.2 and STEP 3.3 on S.

In a nutshell, the set of effective control paths consists of all the distinct sequences of control and expected external actions that can be achieved with a given S and GSO.

Since each effective control path has its last control action corresponding to a goal state, we can define an order relation among the effective control paths in **T** using the '>'-order of the goal states. Although the '>' order is total, the effective control paths are only partially ordered since several effective control paths may have a control action corresponding to the same goal state.

A partially-ordered set of effective control paths is denoted as $\mathbf{T} = <T_1, T_2, ...,$ $T_e>$, i.e. for each consecutive pair T_i, T_{i+1} in T, the goal state of T_{i+1} has no higher priority than the goal state of T_i. The partial priority relation among two effective control paths is denoted as $T_i \geq T_j$.

6.3.3 Example

Figure 25 shows the following states:

$(y_1, u_{1,2})$, $(y_2, u_{2,3})$, $(y_3, u_{3,4})$, $(y_4, u_{4,4})$, $(y_5, u_{5,6})$, $(y_6, u_{6,4})$, $(y_7, u_{7,8})$, $(y_8, u_{8,9})$, $(y_9, u_{9,9})$.

The expected external actions are $u_{2,5}{}^{ext}$, $u_{3,7}{}^{ext}$ and $u_{8,1}{}^{ext}$.

The goal states are $G=\{(y_4, u_{4,4}), (y_9, u_{9,9})\}$ with $(y_4, u_{4,4}) > (y_9, u_{9,9})$.

The set of effective control paths is $T=\{T_1, T_2, T_3\}$, as follows:

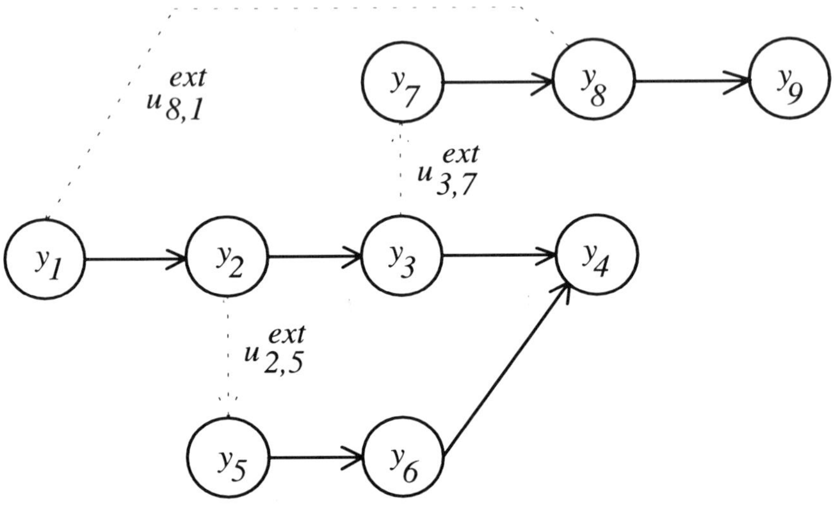

Figure 25 Example to illustrate effective control paths.

$$T_1 = u_{1,2}^{\circ} \Rightarrow u_{2,3}^{\circ} \Rightarrow u_{3,4}^{\circ} \Rightarrow u_{4,4}^{\circ}$$

$$T_2 = u_{1,2}^{\circ} \Rightarrow u_{2,5}^{\circ} \Rightarrow u_{5,6}^{\circ} \Rightarrow u_{6,4}^{\circ} \Rightarrow u_{4,4}^{\circ}$$

$$T_3 = u_{1,2}^{\circ} \Rightarrow u_{2,3}^{\circ} \Rightarrow u_{3,7}^{\circ} \Rightarrow u_{7,8}^{\circ} \Rightarrow u_{8,9}^{\circ} \Rightarrow u_{9,9}^{\circ}$$

In the example above, T_1 has the control action of the optimal goal path towards $(y_4, u_{4,4})$. The effective control path T_2 has control actions and an expected external action corresponding to a goal path towards the same goal, $(y_4, u_{4,4})$ but on a non-optimal goal path. The effective control path T_3 has first the control actions of the states towards $(y_4, u_{4,4})$, aborts this path due to $u_{3,7}^{ext}$ and then follows with the control actions of the states towards the optimal goal state $(y_9, u_{9,9})$. The expected external action $u_{8,1}^{ext}$ is not represented since that would lead to a repetition of the control action $u_{1,2}$.

6.3.4 Partial Effective Control Paths

For a more compact presentation we need a notation for the sequence of actions that starts with the first action symbol in a sequence and ends with a given action symbol in the same sequence.

Definition 6:2. *Partial effective control path.* Let T_n be the effective control path:

$$T_n = u_{1,2}{}^\circ \Rightarrow u_{2,3}{}^\circ \Rightarrow \ldots \Rightarrow u_{i,j}{}^\circ \Rightarrow \ldots \Rightarrow u_{8,8}{}^\circ$$

The *partial effective control path* of T_n relative to $u_{i,j}{}^\circ$, denoted $\Sigma_n(u_{i,j}{}^\circ)$ is the sequence $u_{1,2}{}^\circ \Rightarrow u_{2,3}{}^\circ \Rightarrow \ldots \Rightarrow u_{i,j}{}^\circ$:

$$\Sigma_n(u_{i,j}{}^\circ) = u_{1,2}{}^\circ \Rightarrow u_{2,3}{}^\circ \Rightarrow \ldots \Rightarrow u_{i,j}{}^\circ$$

The index n of Σ is the index of the effective control path T_n.

Example. In the **Figure 25**, the partial effective control path of T_3 relative to $u_{7,8}$ is:

$$\Sigma_3(u_{7,8}) = u_{1,2}{}^\circ \Rightarrow u_{2,3}{}^\circ \Rightarrow u_{3,7}{}^\circ \Rightarrow u_{7,8}{}^\circ$$

Definition 6:3. *Identity relation among partial effective control paths.* Two partial effective control paths are identical if all their action symbols are identical up to and including the action that is the argument for the two partial effective control paths:

Notation: $\Sigma_n(u_{i,j}{}^\circ) \equiv \Sigma_m(u_{i,j}{}^\circ)$.

Example of identical partial effective control path. For the state set in the **Figure 25** we have that:

$$\Sigma_1(u_{2,3}{}^\circ) \equiv \Sigma_3(u_{2,3}{}^\circ) = u_{1,2}{}^\circ \Rightarrow u_{2,3}{}^\circ$$

Some properties of the identity relation among effective control paths.

Since the action symbol of the argument is included in the effective control paths, the identity relation has the following property:

$$\text{IF } \Sigma_n(u_{i,j}{}^\circ) \equiv \Sigma_m(u_{k,l}{}^\circ) \text{ THEN } u_{i,j}{}^\circ \equiv u_{k,l}{}^\circ$$

If $u_{g,g}{}^\circ$ is the last action symbol of the partial effective control path T_m then:

$$T_m \equiv \Sigma_m(u_{g,g}{}^\circ)$$

Definition 6:4. *Inclusion relation among partial effective control paths.* The partial effective control path $\Sigma_n(u_{i,j}{}^\circ)$ is said to be *included* in the partial effective control path $\Sigma_m(u_{k,l}{}^\circ)$ iff $\Sigma_n(u_{i,j}{}^\circ) \equiv \Sigma_m(u_{i,j}{}^\circ)$ and $u_{k,l}{}^\circ \notin \Sigma_m(u_{i,j}{}^\circ)$ (i.e. the action symbol $u_{k,l}{}^\circ$ is placed after $u_{i,j}{}^\circ$ in T_m). Notation: $\Sigma_n(u_{i,j}{}^\circ) \subset \Sigma_m(u_{k,l}{}^\circ)$.

Example.
In **Figure 25**, the partial effective control path $\Sigma_3(u_{2,3}{}^\circ)$ is included in $\Sigma_1(u_{4,4}{}^\circ)$ since $\Sigma_3(u_{2,3}) \equiv \Sigma_1(u_{2,3})$ and $u_{4,4}{}^\circ \notin \Sigma_1(u_{2,3})$

Some properties of the inclusion relation among effective control paths.

If in the definition of the inclusion relation the action $u_{k,l}{}^\circ$ is identical to a control action $u_{g,g}{}^\circ$ which is the last one in the effective control path T_m, then we write that :

$$\Sigma_n(u_{i,j}{}^\circ) \subset T_m$$

The statement that follows is a consequence of how effective control paths are constructed.

For each two arbitrary effective control paths T_n and T_m, $n \neq m$, the following relations holds:

$$T_n \not\subset \Sigma_m(u_{i,j}{}^\circ)$$

$$T_n \not\subset T_m$$

Proof.
Let $u_{g,g}{}^\circ$ be the last symbol in T_n. For the first formula above, if T_n would belong to $\Sigma_m(u_{i,j}{}^\circ)$, this means that $\Sigma_n(u_{g,g}{}^\circ) \equiv \Sigma_m(u_{g,g}{}^\circ)$ and there exists a symbol $u_{g,k}{}^\circ \in T_m$ such that $u_{g,g}{}^c \Rightarrow u_{g,k}{}^\circ$. This implies that the effective control path T_n is not well constructed since the last symbol in T_n has a consecutive action symbol. The proof for the second relation uses the same observation.
Q.E.D.

6.3.5 The Interpretation of Effective Control Paths

The Reason For Interpreting Effective Control Paths

The controller, by means of its GSO has the purpose to materialize goal states in their '>'-order. However, during the control, expected external actions may disrupt the currently followed goal path and the GSO changes to a different optimal or non-optimal goal paths. In terms of effective control paths, this means that the GSO completes states with the external expected and control actions of a particular effective control path T_i in **T,** such that all the expected external actions that occur match the expected external actions in T_i. However, it is not known a priori which one of the effective control paths T_i will be followed since expected external actions occur at unpredictable time instances. Therefore we can find the right T_i by matching the control actions already executed until a time instance with the control actions of some effective control path in **T**. In this section we study the relation between the states completed by the GSO and the corresponding effective control path in **T**.

Informally, this relation is the following.

If all the states in the optimal goal paths materialize in their '>'-order and only those expected external actions occur that override the effect of the control actions of the goal states, then the GSO completes all the control actions of T_1. However, if some expected external action occurs while the controller materializes a state which is not a goal state, then the corresponding control action in T_1 cannot materialize. Yet, due to the way effective control paths are constructed, there exists another effective control path, say T_i, that corresponds to the current control situation and then T_i is followed. Therefore, we can always identify a current effective control path T_i that has the largest partial effective control path whose actions already have been materialized.

For instance, let the first current effective control path be T_1. After an expected external action occurs, the current effective control path may be T_k, $k>1$. After yet another expected external action, which does not appear in T_k, the current effective control path may be $T_i \geq T_k$, and so on. Finally the GSO completes states such that an effective control path T_j is followed completely. The process above shows that we can define the interpretation for an action symbol in an effective control path, for partial effective control paths and for full effective control paths. These interpretations are counterparts for state interpretations and goal paths interpretations and are required later for the definition of proper plant formulas of the so called *GSO-controller*.

The interpretation of one action symbol in an effective control path

Definition 6:5. *Interpretation of action symbols in effective control paths.* Let $(y_i, u_{i,j})$ and $(y_i, u_{i,k}{}^{ext})$ be two states in S. As already defined in Chapter 3, the expected external action $u_{i,k}{}^{ext}$ may occur while the plant output is y_i and it overrides the effect of the control action $u_{i,j}$. Let the state $(y_i, u_{i,j})$ be materialized at t, i.e. $y_i(t)=1$ and $u_{i,j}(t)=1$. At $t' > t$ the interpretation of one action symbol in an effective control path is determined from the pair $(u_{i,j}(t), y_k(t'))$ as follows:

$$\text{If } k = j \text{ then } u_{i,j}(t')=1$$
$$\text{If } k \neq j \text{ then } u_{i,k}{}^{ext}(t')=1$$

The interpretation of a partial effective control path

A partial effective control path is interpreted to 1 at t when all its component action symbols are interpreted to 1 in successive time instances and the last action in it interprets to 1 at t. If at least one action symbol does not interpret to 1 at a time instance, the interpreted value of the effective control path is 0.

Definition 6:6. *The interpretation of a partial effective control path.* Let T_n be the following effective control path:

$$T_n = u_{1,2}{}^{\circ} \Rightarrow u_{2,3}{}^{\circ} \Rightarrow ... \Rightarrow u_{i,j}{}^{\circ} \Rightarrow ... \Rightarrow u_{g,g}{}^{\circ}$$

and

$$\Sigma_n(u_{i,j}{}^{\circ}) = u_{1,2}{}^{\circ} \Rightarrow u_{2,3}{}^{\circ} \Rightarrow ... \Rightarrow u_{i,j}{}^{\circ}$$

be a partial effective control path of T_n. The partial effective control path $\Sigma_n(u_{i,j}{}^{\circ})$ is true at time t iff there exist time instances $t_1 < t_2 < ... < t$ such that:

$$u_{1,2}{}^{\circ}(t_1)=1 \text{ and } u_{2,3}{}^{\circ}(t_2)=1 \text{ and } ...\text{and } u_{i,j}{}^{\circ}(t)=1$$

Notation: $\Sigma_n(u_{i,j}{}^\circ)(t)=1$.

Interpretation of an effective control path

An effective control path is interpreted to 1 at t when all its action symbols interpret to 1 in successive time instances, the last action symbol being interpreted to 1 at time t.

Definition 6:7 *The interpretation of an effective control path.* Let T_n be the following effective control path:

$$T_n = u_{1,2}{}^\circ \Rightarrow u_{2,3}{}^\circ \Rightarrow \ldots \Rightarrow u_{i,j}{}^\circ \Rightarrow \ldots \Rightarrow u_{g,g}{}^\circ$$

The effective control path T_n is true at t iff there exist time instances $t_1 < t_2 < \ldots t_j <\ldots< t$ such that:

$u_{1,2}{}^\circ(t_1)=1$ and $u_{2,3}{}^\circ(t_2)=1$ and ...and $u_{i,j}{}^\circ(t_i)=1$ and ... and $u_{g,g}{}^\circ(t)=1$.

Notation: $T_n(t)=1$.

Commutations among effective control paths

The set of effective control paths was designed such that each sequence of states the GSO can materialize starting from S_1 has its actions in one of the effective control paths T_i. Normally the effective control paths have common partial effective control paths. However it is not known a priori which particular effective control path the GSO will follow since expected external actions may occur at unpredictable time instances. Therefore, any time a control action or an expected external action materializes which is not in the current effective control path, a new effective control path can be determined among those in **T.** The new effective control path contains all the past actions that have been materialized in the old effective control path but also contains the last action symbol which is not in the old effective control path. The operation of changing the current effective control path is called a *commutation*. For the example in **Figure 25**, if the current effective control path is T_1, then a commutation is done from the action symbol $u_{2,3}{}^\circ$ in T_1 to the action symbol $u_{3,7}{}^\circ$ in T_3. The partial effective control path $u_{1,2}{}^\circ \Rightarrow u_{2,3}{}^\circ$ is common to both T_1 and T_2.

The definition of a commutation is the following.

Definition 6:8. *The commutation operation among effective control paths.* Let T_n and T_m be two effective control paths. There exists a *commutation from* $u_{i,j}{}^\circ \in T_n$ *to* $u_{j,l}{}^\circ \in T_m$ $(n \neq m)$ *iff*

$$\Sigma_n(u_{i,j}{}^\circ) \subset \Sigma_m(u_{j,l}{}^\circ) \text{ and } u_{j,l}{}^\circ \notin T_n$$

(Note that $u_{i,j}{}^\circ$ and $u_{j,l}{}^\circ$ have the same j-index. That means T_n includes the transition $u_{i,j}{}^\circ \Rightarrow u_{j,l}{}^\circ$).

The interpretation of an effective control path using commutations

Each inclusion of partial effective control paths, as shown above defines a possible commutation. Therefore, by examining the elements of **T** for inclusions of partial effective control paths, it is possible to detect syntactically the set of all the possible commutations. The last possible commutation occurs when:

$$\Sigma_n(u_{i,g}{}^\circ) \subset \Sigma_m(u_{g,g}{}^\circ) \text{ and } u_{g,g}{}^\circ \notin T_n$$

where $u_{g,g}{}^\circ$ is the last symbol in the sequence T_m. Since there is no partial or total effective control path which includes T_m, there cannot be another commutation starting from T_m.

From the above, it can be seen that an effective control path T_i materializes at t_i as follows. The GSO starts with T_1 and let us assume that a commutation is done at time t_1 from $u_{a,b}{}^\circ \in T_1$ to T_a. From T_a a commutation is done from $u_{c,d}{}^\circ \in T_a$ to T_b, etc. At some time t_n the current effective control path is T_n and a commutation is done at $u_{e,f}{}^\circ \in T_n$. Finally, after a commutation to T_i, at time t_i, the last symbol of T_i materializes. Then, from the definition of the commutation operation and the interpretation of partial effective control paths, we can write the following:

$$\Sigma_1(u_{a,b}{}^\circ) \subset \Sigma_a(u_{c,d}{}^\circ) \subset \dots \subset \Sigma_n(u_{e,f}{}^\circ) \subset \dots \subset T_i$$

and

$$\Sigma_1(u_{a,b}{}^\circ)(t_1) = 1 \text{ and } \Sigma_a(u_{c,d}{}^\circ)(t_a) = 1 \text{ and } \dots \text{and } \Sigma_n(u_{e,f}{}^\circ)(t_n) = 1 \text{ and}\dots\text{and } T_i(t_i)$$
$$= 1$$

It can be seen that by performing commutations, some action symbols in some partial effective control paths are never interpreted. For example, in **Figure 25**, if a commutation is done from $u_{2,3}{}^\circ$ in T_1 to $u_{3,7}{}^\circ$ in T_3 then the actions in the partial effective control path $u_{1,2}{}^\circ \Rightarrow u_{2,3}{}^\circ$ belonging to T_3 will not be interpreted ever, since they are already interpreted at T_1.

Property 6:1. Let there be a commutation from $u_{i,j}{}^\circ \in T_n$ to $u_{j,l}{}^\circ \in T_m$ $(n \neq m)$. Then the partial effective control path $\Sigma_m(u_{i,j}{}^\circ)$ is never interpreted.

This property follows from the definition of the interpretation. The commutation from $u_{i,j}{}^\circ \in T_n$ to $u_{j,l}{}^\circ \in T_m$, according to the definition 6:8 means that:

$$\Sigma_n(u_{i,j}{}^\circ) \subset \Sigma_m(u_{j,l}{}^\circ) \text{ and } u_{j,l}{}^\circ \notin T_n$$

The relation above can be written as:

$$\Sigma_m(u_{j,l}{}^\circ) = \Sigma_n(u_{i,j}{}^\circ) \Rightarrow u_{j,l}{}^\circ.$$

(the notation " $\Sigma_n(u_{i,j}{}^\circ) \Rightarrow u_{j,l}{}^\circ$ " denotes the partial effective control path $u_{1,a}{}^\circ \Rightarrow \dots \Rightarrow u_{i,j}{}^\circ \Rightarrow u_{j,l}{}^\circ$)

The partial effective control path $\Sigma_m(u_{j,l}{}^\circ)$ is also

$$\Sigma_m(u_{j,l}{}^\circ) = \Sigma_m(u_{i,j}{}^\circ) \Rightarrow u_{j,l}{}^\circ$$

However, T_n interprets before T_m and therefore $\Sigma_n(u_{i,j}{}^\circ)$ interprets always before $\Sigma_m(u_{i,j}{}^\circ)$.

The property above means that the partial effective control path $\Sigma_m(u_{i,j}{}^\circ)$ is redundant and it can be removed from T_m.

The interpretation of an effective control path described above has the following consequence.

Property 6:2. There is always a unique effective control path whose control actions are used by the GSO when completing the control actions from an initial state up to a final goal state.

This property will be used in the following sections to define a state set that is different from S but which perform the same control as S.

Conclusions

The description of the operations performed by the GSO in terms of effective control paths, shows that the operation of goal seeking can be considered as a particular case of control. The control goal in this case is to determine which effective control path corresponds to the current control situation. The control actions are commutations among effective control paths. To make the distinction between the controller at the object state level and the controller at the GSO-level, we shall call the latter as the *GSO-controller*.

In the following section we give a formal characterization of the state set of the GSO-controller. This state set is called the *GSO-state set* and is denoted S_{gso}. Then we show that under certain conditions the GSO-state set is well-determined.

6.4 The State Set Of The GSO-controller

In this section we define the state set of the GSO-controller. The control scheme of the GSO-controller, which is adapted after the control scheme from **Figure 9** is shown in the **Figure 26**. The elements of the control scheme are identified in this section together with the components of the GSO-controller state: *GSO-control actions*, *GSO-configurations* and *GSO-proper plant formulas*.

Let S be the state set of a controller with the GSO as defined in Section 3.4.3 and $\mathbf{T} = <T_1, T_2, ..., T_i,...,T_e>$ be the set of effective control paths that can be constructed with S and GSO. The components of the state set of the GSO-controller are the following.

6.4.1 The Configuration Of The GSO-controller

Definition 6:9. *The GSO-configuration.* The configuration of the GSO-controller is the index n of the current effective control path T_n ($n=1,...,e$).

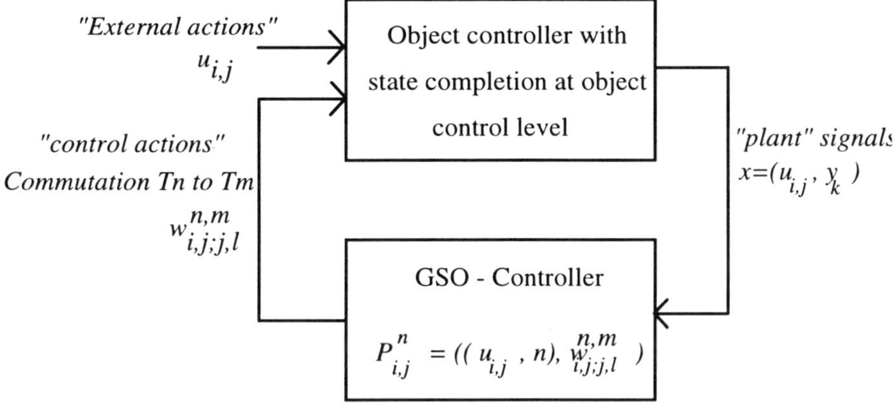

Figure 26 The GSO Controller and its controlled process.

6.4.2 The Control Actions Of The GSO-controller

Definition 6:10. *The GSO-control action.* A GSO-control action relative to two action symbols $u_{i,j}^{\,\circ}$ and $u_{j,k}^{\,\circ}$ is a commutation from $u_{i,j}^{\,\circ}$ of T_n to $u_{j,l}^{\,\circ}$ of T_m as defined by definition 6:8. Notation: $w_{i,j;j,l}^{n,m}$.

In other words a GSO-control action is defined relative to two action symbols $u_{i,j}^{\,\circ}$ and $u_{j,l}^{\,\circ}$, that have their partial effective control paths $\Sigma_n(u_{i,j}^{\,\circ}) \subset \Sigma_m(u_{j,l}^{\,\circ})$, with $n \neq m$.

6.4.3 The Proper Precondition Formula Of The GSO-controller

Definition 6:11. *The proper precondition formula of the GSO-controller.* The proper precondition formula of the GSO-controller is an action symbol $u_{i,j}^{\,\circ}$ materialized according to the definition 6:5.

6.4.4 The State Set Of The GSO-controller.

Definition 6:12. *A GSO-state.* A GSO-state is a tuple:

$$((u_{i,j}^{\,\circ}, n), w_{i,j;j,l}^{n,m})$$

where:

$u_{i,j}{}^{\circ}$ is the GSO-proper precondition formula according to the definition 6:11

n is the GSO-configuration according to the definition 6:9

$w_{i,j;j,l}{}^{n,m}$ is the GSO-control action according to the definition 6:10 and $n \neq m$.
The condition on the indices i,j,k is that i, j, l are distinct, or either $i \equiv j$ or $j \equiv l$ but not both.

We use the notation $P_{i,j}{}^{n}$ for a GSO-state $((u_{i,j}{}^{\circ}, n), w_{i,j;j,l}{}^{n,m})$.

A goal state of the GSO-controller, called GSO-goal state, is represented using the same convention as in Section 4.6, i.e. the GSO-control action of this state does not change the GSO-configuration, nor the GSO-proper postcondition:

Definition 6:13. *GSO-goal state.* A GSO-goal state is defined as:

$$P_{i,j}{}^{n}=((u_{i,i}{}^{\circ}, n), w_{i,i;i,i}{}^{n,n})$$

where $w_{i,i;i,i}{}^{n,n}$ denotes that the control action $u_{i,i}$ is true at t and at any time instance $t' > t$ and $u_{i,i}{}^{\circ}$ is the last action symbol in the effective control path T_n.

This definition fulfills the condition required for those goal states whose control action is the last in an effective control path. The condition is that no expected external action can occur from that particular goal state. Therefore the action $u_{i,i}$ materializes indefinitely.

Definition 6:14 *The GSO-state set.* The GSO-state set is the set:

$$P=\{((u_{i,j}{}^{\circ}, n), w_{i,j;j,l}{}^{p,q})\}$$

where i, j ranges over the number of plant formulas and p, q ranges over the number of effective control paths, $u_{i,j}{}^{\circ}$ is a GSO-proper precondition formula according to definition 6:11, p is the GSO-configuration according to the definition 6:9 and $w_{i,j;j,l}{}^{p,q}$ is the GSO-control action according to the definition 6:10. When $i \equiv j$ then $p \equiv q$ and the state is a goal state.

6.4.5 GSO-state Transitions

According to what was described above, the GSO-state transition is a commutation of the effective control path. Let there be two states in the GSO-state set:

$$P_{i,j}{}^n = ((u_{i,j}{}^\circ, n), w_{i,j;j,k}{}^{n,m})$$

$$P_{j,k}{}^m = ((u_{j,k}{}^\circ, m), w_{j,k;k,l}{}^{m,r})$$

Definition 6:15. *The GSO-state transition.* A GSO-state transition is $P_{i,j}{}^n \to P_{j,k}{}^m$:

$$((u_{i,j}{}^\circ, n), w_{i,j;j,k}{}^{n,m}) \to ((u_{j,k}{}^\circ, m), w_{j,k;k,l}{}^{m,r})$$

takes place when $P_{i,j}{}^n$ materializes at t and $P_{j,k}{}^m$ materializes at some time $t' > t$.

Considering the definition for all the components of the GSO-states, the definition above has the following meaning.

At time t, the control action $u_{i,j}$ or the expected external action $u_{i,j}{}^{ext}$ materializes and the corresponding symbol $u_{i,j}{}^\circ$ belongs to the effective control path T_n. At time $t' > t$, the control action $u_{j,k}$ or the expected external action $u_{j,k}{}^{ext}$ materializes such that $u_{j,k}{}^\circ$ does not belong to T_n but it belongs to T_m. That is an indication that control actions from T_n cannot be completed any more but the completion operation completes control actions in the effective control path T_m.

6.4.6 The Expected External Actions Of The GSO-controller

The occurrence of a GSO-expected external action simply means that the plant formula of the next state in the current effective control path materializes as expected and no commutation of the current effective control path is required. The expected external action for the GSO-controller has the same meaning as for the object controller (see Section 4.1.6), i.e. the expected external action occurs at some unpredictable time and it does not change the controller configuration. The unpredictability of the time is due to the fact that a number of inner state transitions may occur until the control action of a consecutive state materializes. To distinguish an expected external action of the GSO from the expected external action of the object controller, we call the former a *GSO-expected external action.*

Thus a GSO-expected external action is defined by a transition $u_{i,j}{}^{\circ} \Rightarrow u_{j,k}{}^{\circ}$ belonging to an effective control path. However, the transition $u_{i,j}{}^{\circ} \Rightarrow u_{j,k}{}^{\circ}$ may belong to several effective control paths. The definition of the commutation among effective control paths shows that only those partial effective control paths can materialize which are not excluded by the property 6:1. More precisely, the definition of a GSO-control action $w_{i,j;j,l}{}^{n,m}$ implies $\Sigma_n(u_{i,j}{}^{\circ}) \subset \Sigma_m(u_{j,l}{}^{\circ})$ and that means $\Sigma_m(u_{j,j}{}^{\circ})$ is never interpreted since $\Sigma_n(u_{i,j}{}^{\circ}) \equiv \Sigma_m(u_{j,j}{}^{\circ})$ and the partial effective control path $\Sigma_n(u_{i,j}{}^{\circ})$ interprets before $\Sigma_m(u_{j,j}{}^{\circ})$. Thus the GSO-expected external action is defined as follows:

Definition 6:16. *GSO-expected external action.* A GSO-expected external action $w_{i,j;j,l}{}^{ext}$ exists in S_{gso} iff

- There exists a T_n such that $\Sigma_n(u_{i,j}{}^{\circ}) \subset \Sigma_n(u_{j,l}{}^{\circ})$

- There is no GSO-expected external action $w_{p,q;r,s}{}^{m,n}$ such that
$\Sigma_n(u_{j,l}{}^{\circ}) \subset \Sigma_n(u_{r,s}{}^{\circ})$

In other words the first condition ensures that an external action $w_{i,j;j,l}{}^{ext}$ exists while the second condition ensures that it is exactly under the configuration T_n when the GSO-expected external action may materialize.

A state with GSO-expected external action is denoted $P_{i,j}{}^{n}=((u_{i,j}{}^{\circ}, n), w_{i,j;j,l}{}^{ext})$.

A GSO-expected external action, denoted $w_{i,j;j,l}{}^{ext}$, is known to occur at t, whenever the GSO-state materialized at t is $((u_{i,j}{}^{\circ}, n), w_{i,j;j,k}{}^{n,m})$ and the GSO-plant formula pair materialized at $t' > t$ is $(u_{j,l}{}^{\circ}, n)$.

6.4.7 Notation Cross-reference

The state set of the GSO is a state with configurations identical to the state set defined in Section 4.3. To make a difference between the two state types, we used for the definitions 6:11 to 6:15 a different notation. A cross-reference between the two state notations is shown in Table 4 below.

Table 4

State Component	Object Control State	GSO State
Proper plant formula	z_i	$u_{i,j}{}^{\circ}$
Configuration	c_i	*<eff. control path index>*
Control action	$u_{i,j}{}^{p,q}$	$w_{i,j;j,k}{}^{p,q}$
Expexted external action	$u_{i,j}{}^{ext}$	$w_{i,j;j,k}{}^{ext}$
State	$S_i{}^{n}=((z_i,\,c_i),\,u_{i,j}{}^{n,m})$	$P_{i,j}{}^{n}=((u_{i,j}{}^{\circ},\,n),\,w_{i,j;j,k}{}^{n,m})$
Goal state	$G_i{}^{n}=((z_i,\,c_i),\,u_{i,i}{}^{n,n})$	$P_{i,i}{}^{n}=((u_{i,i}{}^{\circ},\,n),\,w_{i,i;i,i}{}^{n,n})$

6.4.8 How to Construct The GSO-state Set

The construction of the GSO-state set using the state set of the object controller can be done using the three phases below using a simple graphical representation. The graphical representation uses nodes for plant formulas and arcs for control and expected external actions.

Phase 1

Using the state set S of the object controller and the expected external actions, in this phase we determine all the possible effective control paths T_i according to the definition 6:1. Then, the effective control paths can be ordered (partially) after the priority of their last goal state. Graphically, the effective control paths can be drawn as nodes representing the plant formulas $u_{i,j}{}^{\circ}$ that are linked by arcs according to the order the control actions and expected external actions appear in the current effective control path.

Phase 2

In this phase, the GSO-control actions are determined according to the definition 6:10. In a graphical representation, the GSO-control actions are bold arcs between GSO-plant formulas belonging to distinct effective control paths. If a commutation is done such that $\Sigma_n(u_{i,j}{}^{\circ}) \subset \Sigma_m(u_{i,l}{}^{\circ})$, then the partial effective control path $\Sigma_m(u_{i,j}{}^{\circ})$ can be removed since it cannot be reached. Graphically, if a bold arc reaches the node $u_{i,j}{}^{\circ}$, all the nodes and links before $u_{i,j}{}^{\circ}$ shall be removed.

Phase 3

In this phase we determine the GSO-states. The bold arcs are GSO-control actions and the arcs in the effective control paths are GSO-expected external actions.

Example

For the example in **Figure 25**, the three phases are shown below.

Phase 1: the set of effective control paths

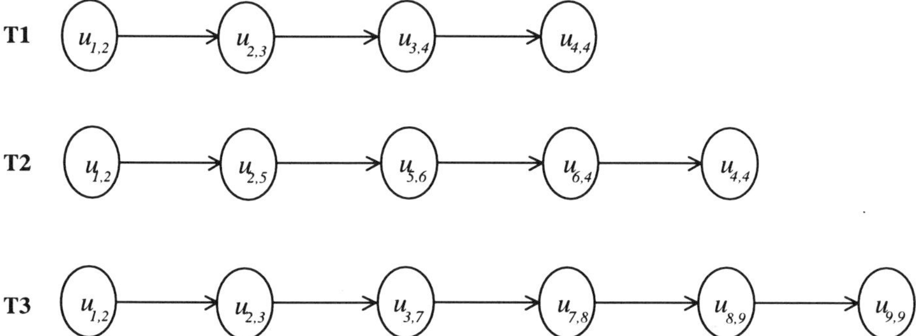

Figure 27 The construction of a GSO-state set, phase 1.

Phase 2: GSO-control actions

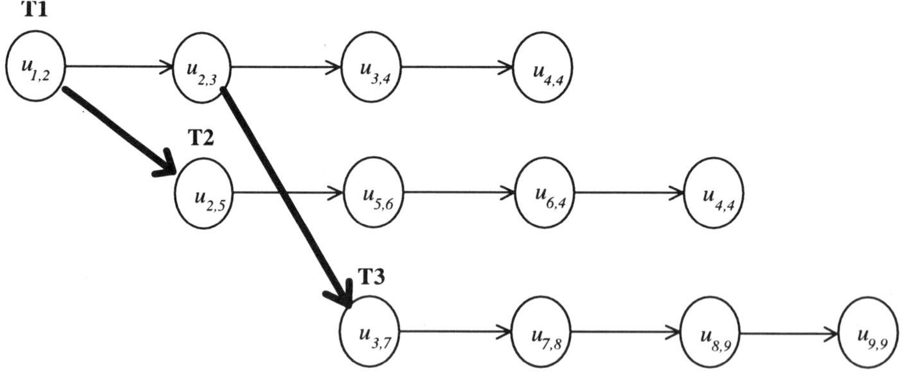

Figure 28 The construction of a GSO-state set, phase 2.

In the **Figure 28** above the GSO-control actions are drawn with bold arrows and are $w_{1,2;2,5}^{1,2}$ and $w_{2,3;3,7}^{1,3}$. The arrows that are not bold are GSO-expected external actions.

Phase 3: the GSO-state set.

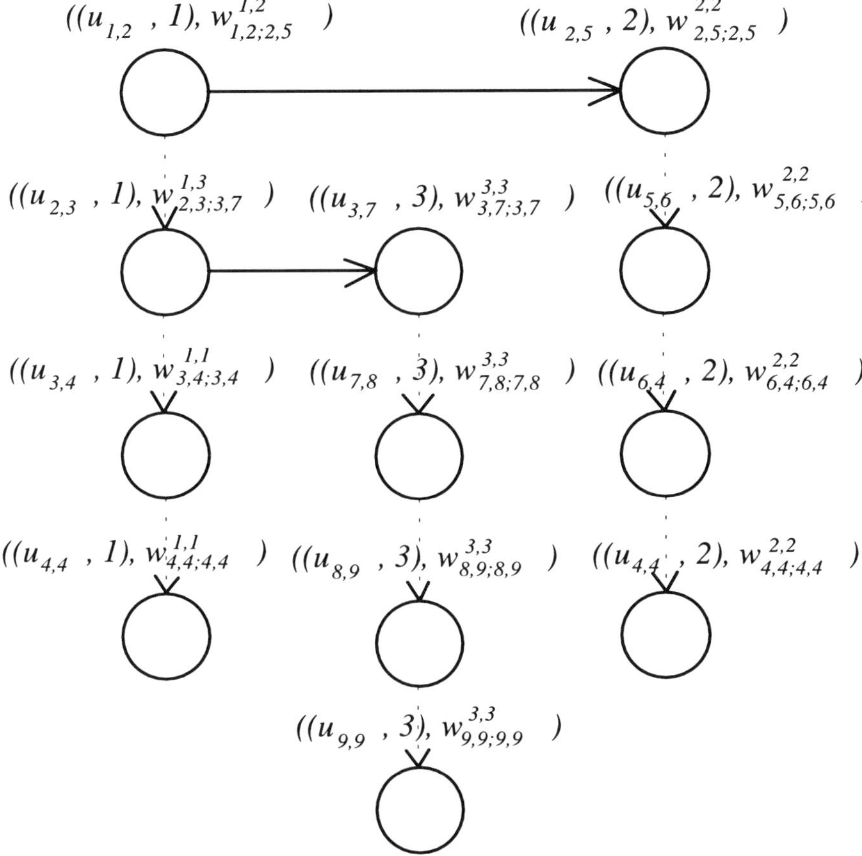

Figure 29 Construction of the GSO-state set, phase 3.

We remark that the goal states of the GSO-controller are $((u_{3,4}°, 1), w_{3,4;3,4}^{1,1})$, $((u_{2,5}°, 2), w_{2,5;2,5}^{2,2})$ and $((u_{3,7}°, 3), w_{3,7;3,7}^{3,3})$ since at these states the effective control path is determined and there are no further control actions that can change the current effective control path. However, from these goal states, several expected external actions may materialize further states. The states that have no further expected external actions are $((u_{4,4}°, \ 1), \ w_{4,4;4,4}{}^{1,1})$ and $((u_{9,9}°, \ 3),$ $w_{9,9;9,9}{}^{3,3})$.

6.5 A Well Determined GSO-State Set

In this section we show that the GSO-state, under certain conditions, is a well-determined state set. According to the definition in Section 6.5.5, the GSO-state set is well-determined if it has the following properties:

- S has the control integrity property (defined in Section 4.5.1)

- S has the specificity of control configuration property (defined in Section 4.5.2)

- S has the specificity of control action property (defined in Section 4.5.3)

- S is syntactically complete (defined in Section 4.5.4)

At the end of this section we show that the conditions imposed on the state set S_{gso} such that it is well determined, are sound (from a process control perspective) for a carefully designed control system and can be achieved with a certain modification of the synchronization step of the GSO.

6.5.1 The Control Integrity Property For S_{gso}

The state set S_{gso} satisfies the control integrity property without any auxiliary restrictions as shown below.

Proposition 6:1. The state set S_{gso} has the control integrity property.

Using the notations for the GSO-states, the control integrity property 4.5.1, is the following:

The state set Sgso contains the state:

$$P_{i,j}{}^n = ((u_{i,j}°, n), w_{i,j;j,j}{}^{n,m})$$

or the state,

$$P_{k,j}{}^m = ((u_{k,j}°, m), w_{k,j;j,j}{}^{ext})$$

but not both states.

Proof.

Let us assume that both states exist. The definition of the GSO-control action gives for $P_{i,j}{}^n$:

$$\Sigma_n(u_{i,j}{}^\circ) \subset \Sigma_m(u_{j,l}{}^\circ)$$

The definition of the GSO-expected external action gives for $P_{k,j}{}^m$:

$$\Sigma_m(u_{k,j}{}^\circ) \subset \Sigma_m(u_{j,l}{}^\circ) \qquad\qquad (*)$$

The first inclusion can be written as:

$$\Sigma_m(u_{j,l}{}^\circ) \equiv \Sigma_n(u_{i,j}{}^\circ) \Rightarrow u_{j,l}{}^\circ$$

The second inclusion can be written as:

$$\Sigma_m(u_{j,l}{}^\circ) \equiv \Sigma_n(u_{k,j}{}^\circ) \Rightarrow u_{j,l}{}^\circ$$

The last two equations imply that:

$$\Sigma_n(u_{i,j}{}^\circ) \equiv \Sigma_n(u_{k,j}{}^\circ)$$

According to the property of partial effective control paths, the last equation implies $u_{i,j}{}^\circ \equiv u_{k,j}{}^\circ$. Replacing $u_{k,j}{}^\circ$ with $u_{i,j}{}^\circ$ in the equation $(*)$ we have that:

$$\Sigma_m(u_{i,j}{}^\circ) \subset \Sigma_m(u_{j,l}{}^\circ)$$

This equation contradicts the second required condition for GSO-expected external actions from definition 18:5.

Q.E.D.

6.5.2 The Specificity Of Control Configuration For S_{gso}

The property of specificity of control configuration 20:5 states that if two states are both consecutive to states with control actions, then their configuration must be different. In terms of GSO-states, that means no two GSO-control actions have their second action symbol belonging to the same effective control path. This property holds for S_{gso} without other preconditions.

Proposition 6:2. *There cannot be two states $P_{i,j}{}^n$ and $P_{q,r}{}^v$ in S_{gso} such that:*

$$P_{i,j}{}^n = ((u_{i,j}{}^\circ, n), w_{i,j;j,k}{}^{n,m})$$

and

$$P_{q,r}{}^v = ((u_{q,r}{}^\circ, v), w_{q,r;r,s}{}^{v,m}).$$

(Note that the two states $P_{i,j}{}^n$ and $P_{q,r}{}^v$ have consecutive states with the same configuration m).

Proof.
The definition for the GSO-control action 6:10 implies that $u_{j,k}{}^\circ \in T_m$ (for the GSO-state $P_{i,j}{}^n$) and that $u_{r,s}{}^\circ \in T_m$ (for the GSO-state $P_{q,r}{}^v$). Without loss of generality, let us assume that $u_{r,s}{}^\circ$ comes before $u_{j,k}{}^\circ$ in T_m. Then $\Sigma_m(u_{r,s}{}^\circ) \subset \Sigma_m(u_{j,k}{}^\circ)$. However, according to the property 6:1, the partial effective control path $\Sigma_m(u_{r,s}{}^\circ)$ does not appear in T_m.
Q.E.D.

6.5.3 The Specificity Of Control Action For S_{gso}

The specificity of control action (Section 4.5.3) requires that there are no two GSO-states in S_{gso} which have the same precondition pair but different GSO-control actions. The specificity of control action is not inherent in the properties of the S_{gso} state set. For example each two states at the object control level that have the same plant formula and two different expected external actions, result in two distinct GSO-control actions that have the same GSO-precondition and thus the property of specificity of control action is not satisfied. However, if the set of effective control paths **T** fulfills certain constraints, the state set S_{gso} can be shown to have the property of specificity of control action. In Section 6.5.5 we discuss the meaning of these constraints and we show that these are sound from a process control perspective.
First we define the constraints in **T**.

Let **T** be a set of effective control paths corresponding to a state set S and a GSO. The set **T** is said to be *distinct* iff for each two action symbols:

$$u_{a,j}{}^{\circ} \in T_m \text{ and } u_{b,j}{}^{\circ} \in T_p$$

we have that:

IF $\qquad ((u_{a,j}{}^{\circ} = u_{a,j} \text{ and } u_{b,j}{}^{\circ} = u_{b,j})$ or

$\qquad\qquad (u_{a,j}{}^{\circ} = u_{a,j}{}^{ext} \text{ and } u_{b,j}{}^{\circ} = u_{b,j}{}^{ext}))$ and if $a \neq b$

THEN $\qquad \Sigma_m(u_{a,j}{}^{\circ}) \neq \Sigma_p(u_{b,j}{}^{\circ}).$

In other words, the requirement is that for each two actions of the same type (i.e. both are control actions or both are expected external actions) and which belong to object states with the same postcondition (i.e. both object states to which $u_{a,j}{}^{\circ}$ and $u_{b,j}{}^{\circ}$ belong have the postcondition y_j), the partial effective control paths of the two action symbols should be distinct.

Proposition 6:3. *Let S_{gso} be a state set defined on a set of effective control paths T that is distinct. Then S_{gso} has the property of specificity of GSO-control action.*

Proof.

Let $u_{i,j}{}^{\circ}$ be an arbitrary action symbol in T_n. We have to show that under the conditions above there are no two states $P_{i,j}{}^n = ((u_{i,j}{}^{\circ}, n), w_{i,j;j,k}{}^{n,m})$ and $P_{i,j}{}^n = ((u_{i,j}{}^{\circ}, n), w_{i,j;j,l}{}^{n,p})$ in S_{gso} that have the same precondition $(u_{i,j}{}^{\circ}, n)$ and different GSO-control actions $w_{i,j;j,k}{}^{n,m} \neq w_{i,j;j,l}{}^{n,p}$.

The definition 6:10 of the GSO-control action implies that there exist two action symbols $u_{j,k}{}^{\circ} \in T_m$ and $u_{j,l}{}^{\circ} \in T_p$ to which the commutation from $u_{i,j}{}^{\circ}$ is performed. Both control actions have $u_{i,j}{}^{\circ}$ as their predecessor in T_m, respectively T_p. The condition for the two GSO-control actions is:

$$\Sigma_m(u_{i,j}{}^{\circ}) \subset \Sigma_n(u_{j,k}{}^{\circ})$$
$$\Sigma_p(u_{i,j}{}^{\circ}) \subset \Sigma_n(u_{j,l}{}^{\circ})$$

The condition above implies that:

$$\Sigma_m(u_{i,j}{}^{\circ}) \equiv \Sigma_n(u_{i,j}{}^{\circ})$$
$$\Sigma_p(u_{i,j}{}^{\circ}) \equiv \Sigma_n(u_{i,j}{}^{\circ})$$

The two conditions above also imply that $\Sigma_m(u_{i,j}{}^\circ) \equiv \Sigma_p(u_{i,j}{}^\circ)$. We can distinguish the following cases, depending on the type of the action symbols $u_{j,k}{}^\circ$ and $u_{j,l}{}^\circ$:

Case 1: $u_{j,k}{}^\circ = u_{j,k}$ and $u_{j,l}{}^\circ = u_{j,l}$. Since $\mathbf{T}$ is *distinct* $\Sigma_m(u_{i,j}{}^\circ) \neq \Sigma_p(u_{i,j}{}^\circ)$ and either $w_{i,j;j,k}{}^{n,m}$ exists or $w_{i,j;j,l}{}^{n,p}$ exists but not both.

Case 2: $u_{j,k}{}^\circ = u_{j,k}$ and $u_{j,l}{}^\circ = u_{j,l}{}^{ext}$. Since $\mathbf{T}$ is *distinct* $\Sigma_m(u_{i,j}{}^\circ) \neq \Sigma_n(u_{i,j}{}^\circ)$ and $w_{i,j;j,k}{}^{n,m}$ cannot exist.

Case 3: $u_{j,k}{}^\circ = u_{j,k}{}^{ext}$ and $u_{j,l}{}^\circ = u_{j,l}$. Since $\mathbf{T}$ is *distinct* $\Sigma_p(u_{i,j}{}^\circ) \neq \Sigma_n(u_{i,j}{}^\circ)$ and $w_{i,j;j,l}{}^{n,p}$ cannot exist.

Case 4: $u_{j,k}{}^\circ = u_{j,k}{}^{ext}$ and $u_{j,l}{}^\circ = u_{j,l}{}^{ext}$. Since $\mathbf{T}$ is *distinct* $\Sigma_m(u_{i,j}{}^\circ) \neq \Sigma_b(u_{i,j}{}^\circ)$ and either $w_{i,j;j,k}{}^{n,m}$ exists or $w_{i,j;j,l}{}^{n,p}$ exists but not both.

The four cases above cover all the possible combinations of object-level action types (i.e. expected external actions and control actions). In none of these four cases is possible that both the GSO-control actions $w_{i,j;j,k}{}^{n,m}$ and $w_{i,j;j,l}{}^{n,p}$ can exist.
Q.E.D.

6.5.4 Syntactical Completeness For S_{gso}

The property of syntactical completeness for the GSO-state set is the following:

Proposition 6:4. The state set S_{gso} is syntactically complete iff:

- For each GSO-proper plant formula $u_{i,j}{}^\circ$ there exists at least one configuration n and a control action $w_{i,j;j,l}{}^{n,m}$ in some state of S_{gso}.

- For each GSO-expected external action $w_{i,j;j,l}{}^{ext}$ there exist exactly two proper plant formulas $u_{i,j}{}^\circ$ and $u_{j,l}{}^\circ$ in some states of S_{gso}.

- For each control GSO-control action $w_{i,j;j,l}{}^{n,m}$ there exist at least one GSO-plant formula $u_{i,j}{}^\circ$ in some state of S_{gso}.

The property of syntactical completeness is true due to the way the GSO state set is constructed: (1) each $u_{i,j}{}^\circ$ belong to some effective control path T_n, (2) the existence of $u_{i,j}{}^\circ$ and $u_{j,l}{}^\circ$ for $w_{i,j;j,l}{}^{ext}$ is ensured by the definition of the GSO-expected external action and (3) the existence of $u_{i,j}{}^\circ$ for $w_{i,j;j,l}{}^{n,m}$ is guaranteed by the definition of a GSO-control action.

6.5.5 Well-determined GSO-state set

Proposition 6:5. A GSO-state set as defined in 6:14 on a set of distinct, effective control paths **T**, is well-determined.

Proof.
Follows from 6:1, 6:2, 6:3 and 6:4.
Q.E.D.

From the perspective of process control, the condition on the effective control path **T** to be *distinct* has the following significance:

1. Significance for control actions. Let two states at the object control level be $(y_j, u_{j,k})$ and $(y_j, u_{j,l})$ and let the actions executed prior to $u_{j,k}$ and $u_{j,l}$ be respectively $u_{a,j}{}^\circ$ and $u_{b,j}{}^\circ$. The property of **T** being *distinct* requires that when two different control actions $u_{j,k}$ and $u_{j,l}$ belong to two states with the same formula y_j, then there should be a difference in the past transitions for the two states. This difference is used to distinguish which one of the two control actions to be executed. The condition that expresses this difference in the past transitions is that $\Sigma_n(u_{a,j}{}^\circ) \neq \Sigma_m(u_{b,j}{}^\circ)$. If y_j would be reachable only via a unique sequence of state transitions and thus there is no distinction in the past state transitions of $(y_j, u_{j,k})$ and $(y_j, u_{j,l})$, then there is no available criteria to distinguish which one of the two control actions $u_{j,k}$ and $u_{j,l}$ shall be executed.

2. Significance for expected external actions. Let two states at object control level be $(y_j, u_{j,k}{}^{ext})$ and $(y_j, u_{j,l}{}^{ext})$ and let the actions executed prior to $u_{j,k}{}^{ext}$ and $u_{j,l}{}^{ext}$ be $u_{a,j}{}^\circ$ and respectively $u_{b,j}{}^\circ$. The property of **T** being *distinct* requires again that there should be a difference between the past state transitions of $(y_j, u_{j,k}{}^{ext})$ and $(y_j, u_{j,l}{}^{ext})$, expressed by $\Sigma_n(u_{a,j}{}^\circ) \neq \Sigma_m(u_{b,j}{}^\circ)$. The process control argument for this condition is the following. Since the external actions $u_{j,k}{}^{ext}$ and $u_{j,l}{}^{ext}$ may occur at any time instance, it is possible that they may occur at the same time instance. The control algorithm should have a priority for which one of the effects of the two expected external actions should be considered first. The implementation of such a priority relation implies that

there cannot be the same y_j in both $(y_j, u_{j,k}{}^{ext})$ and $(y_j, u_{j,l}{}^{ext})$ since y_j in one of the states has a priority relation while the other one has another priority relation. That implies also $\Sigma_n(u_{a,j}{}^\circ) \neq \Sigma_m(u_{b,j}{}^\circ)$.

The set of effective control paths **T** can be made *distinct* using a modification of the GSO as follows: the extended image state $<(y_i, u_{i,j})$; PGS ; OGS ; NOGS ; '>' > defined in Section 3.4.2 is amended with a set of expected external actions called EEA and thus is defined as: $<(y_i, u_{i,j})$; PGS ; OGS ; NOGS ; EEA, '>' >. The set EEA contains those expected external actions which make the required distinction among effective control paths as explained under the points (1) and (2) above. Thus each two states that have their y_i, PGS, OGS and NOGS identical, should have their EEA member distinct. In this way, the set of effective control paths that is constructed with the amended GSO is *distinct*.

6.5.6 Example 2

The example below illustrates the construction of a well-determined state set on a more complex example. The state set represented in the **Figure 30** is clearly not well determined: for instance the states $(y_9, u_{9,10})$, $(y_9, u_{9,11})$, $(y_9, u_{9,12})$ do not satisfy the specificity of control action (5.5.3) and the states $(y_{13}, u_{13,14})$ and $(y_9, u_{9,14}{}^{ext})$ do not satisfy the integrity property (5.5.1). However, the effective control paths $T_1,..., T_7$ satisfy all the required properties such that the transformed state set is well-determined. The example uses the modification of the extended state set described earlier. For instance, when the GSO completes the partial state y_9, then:

- if the EEA has $\{u_{2,4}{}^{ext}\}$ then it completes $u_{9,10}$

- if the EEA has $\{u_{5,9}{}^{ext}\}$ then it completes $u_{9,11}$

- if the EEA has $\{u_{4,13}{}^{ext}, u_{13,9}{}^{ext}\}$ then it completes $u_{9,12}$

The details of the construction can be followed in the figures below.

The state set is:

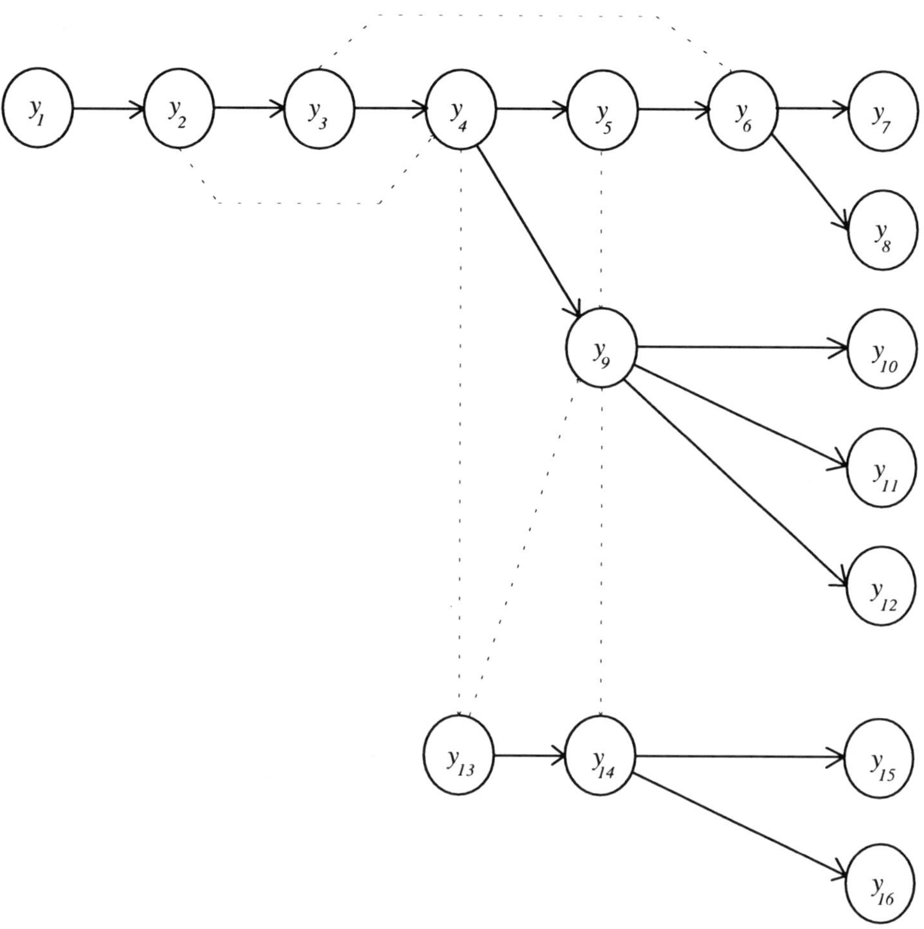

Figure 30 Example 2: State Set.

The effective control paths are:

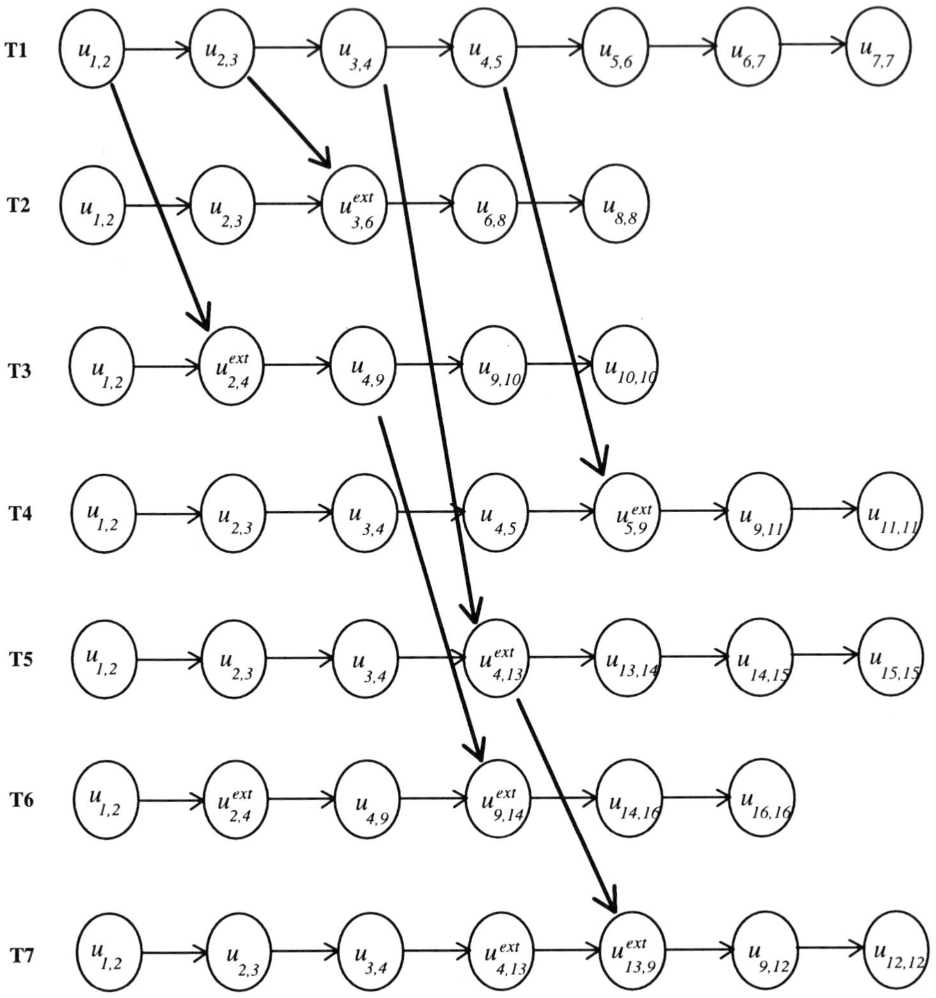

Figure 31 Effective control paths for example 2.

The resulting GSO-state set is:

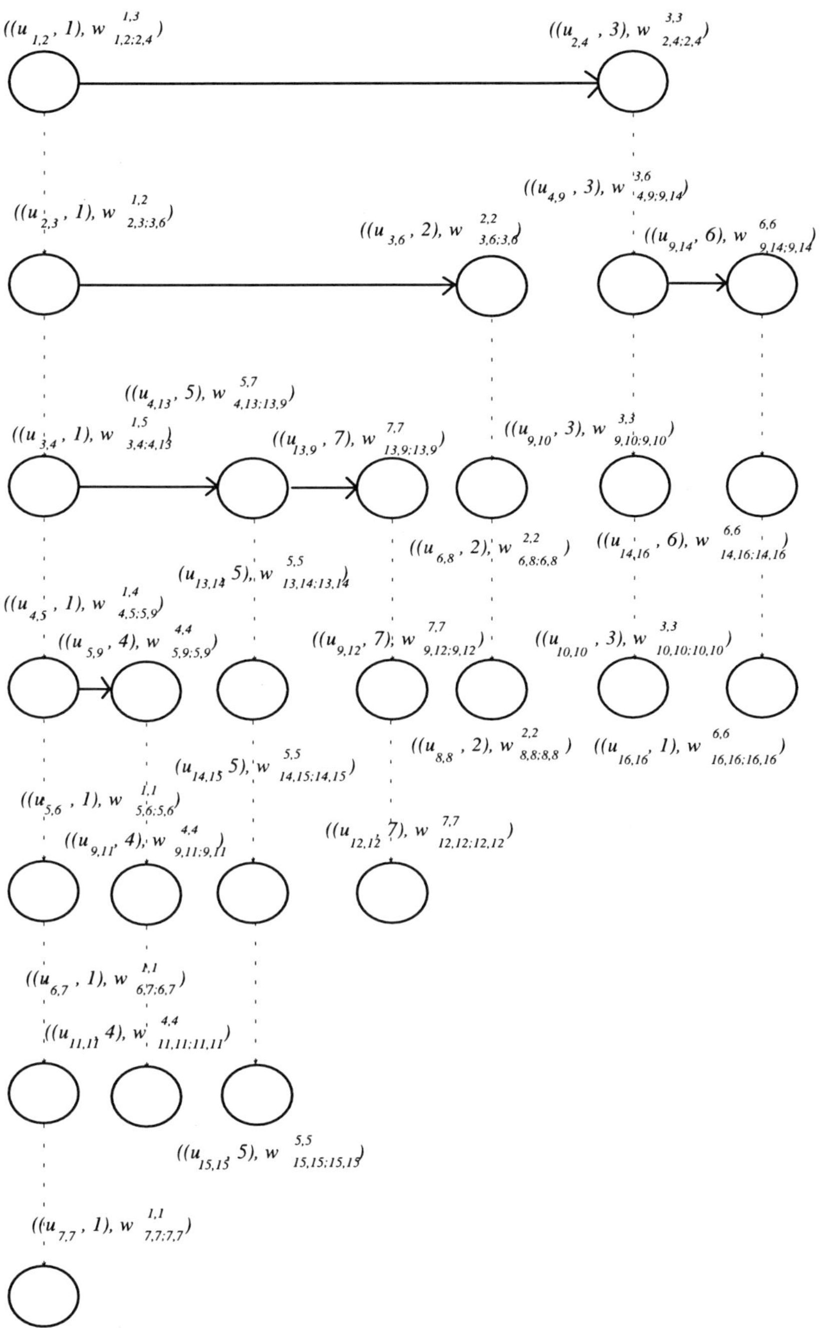

Figure 32 Resulting GSO-state set for example 2.

6.6 De-synchronizations With The GSO-Controller

In Section 6.5 we showed that under certain conditions, we can construct a well-determined state set S_{gso} with a non well-determined state set S and a goal seeking operation GSO. Since S_{gso} is well-determined, we can now apply the goal seeking operation GSO_{wds}. By design, a controller using the pair (S_{gso}, GSO_{wds}) acts identically as the controller using the pair (S, GSO) in cases when (*i*) control actions have expected effects and (*ii*) when expected external actions occur. The explanation is that the controller with the pair (S_{gso}, GSO_{wds}) follows a unique effective control path T_n and T_n consists on the given expected external actions and the control actions the GSO would complete using states in S.

Moreover, since S_{gso} is a well-determined state set, the controller using the pair (S_{gso}, GSO_{wds}) exhibits a characteristic behavior in each one of the four cases of de-synchronization described in Sections 6.3.1, 6.3.2, 6.3.3, 6.3.4 (summary in Table 3). However, this characteristic behavior is relative to de-synchronizations at the abstraction level of the state set S_{gso}, which, as already explained, is different from the abstraction level of the state set S. Still, one can reasonable expect that a certain de-synchronization type in S gives the same de-synchronization type in S_{gso} since (S, GSO) and (S_{gso}, GSO_{wds}) are formal representations derived from the same control knowledge.

The remaining part of Chapter 5 describes this relation.

6.6.1 Ontological De-synchronizations

According to the results from Section 5.1, if a state set of an object controller is well-determined, then VOA can be detected by the existence of an *ontological de-synchronization*, i.e. a transition from a state S_i to $K_-(S_j)$, where S_j is the consecutive state so S_i. Since the state set S we consider in this chapter is not well-determined, then according to the main result from Chapter 3, we have no formal representation for how a controller using S behaves under VOA. Therefore we cannot make a direct proof that each VOA on S means a VOA on S_{gso}. However, we have the indirect proof that the controller using (S_{gso}, GSO_{wds}) will detect any VOA.

Still, for the particular case when a VOA on S does occur as an *ontological de-synchronization*, one should expect that an ontological de-synchronization will occur for S_{gso} as well. The following holds:

Let S be a state set that is not necessarily well-determined and let S_{gso} be a well-determined state set constructed with a *distinct* set **T**. Then we have the following proposition:

Proposition 6:6. An ontological de-synchronization in (S, GSO) is an ontological de-synchronization in (S_{gso}, GSO_{wds}).

Proof.

Let there be the following states in S: $(y_h, u_{h,i})$, $(y_i, u_{i,j})$, $(y_i, u_{i,k}^{ext})$ and $(y_k, u_{k,l})$ (**Figure 33**):

The effective control paths that include the control actions of these states are T_n and T_m:

$$T_n = ... \, u_{h,i}{}^\circ \Rightarrow u_{i,j}{}^\circ \Rightarrow ....$$
$$T_m = ... \, u_{h,i}{}^\circ \Rightarrow u_{i,k}{}^\circ \Rightarrow u_{k,l}{}^\circ \Rightarrow ....$$

If a VOA on (S, GSO) occurs as an ontological de-synchronization, we have that at t $y_h(t) = 1$, and at $t' > t$ $y_k(t') = 1$ such that $(y_k, u_{k,l}) \in K_-(y_i, u_{i,j})$ (see Section 5.3.1).

For S_{gso} there exists a GSO-control action $w_{h,i;i,k}{}^{n,m}$ since according to definition 6:10:

$$\Sigma_n(u_{h,i}{}^\circ) \subset \Sigma_m(u_{i,k}{}^\circ) \text{ and } u_{i,k}{}^\circ \notin T_n.$$

The GSO-postcondition of the GSO-state $P_{h,i}{}^n = ((u_{h,i}{}^\circ, n), w_{h,i;i,k}{}^{nm})$ is $(u_{i,k}{}^\circ, m)$ and the GSO-postcondition due to a GSO-expected external action from a state $((u_{i,k}{}^\circ, m), \_)$ is $(u_{k,l}{}^\circ, m)$.

We can compare now how states materialize for the two controllers:

(S, GSO)

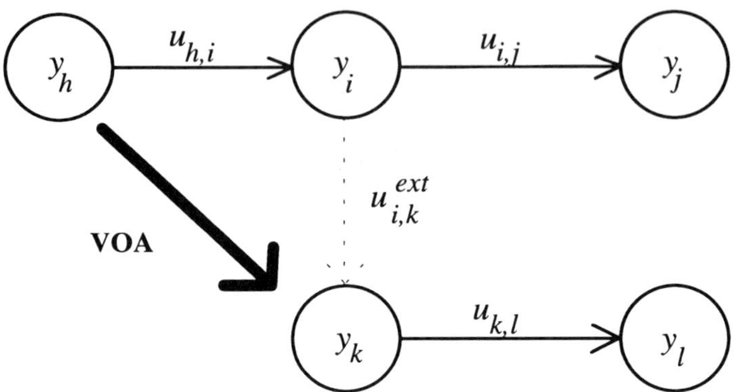

(Sgso, GSOwds)

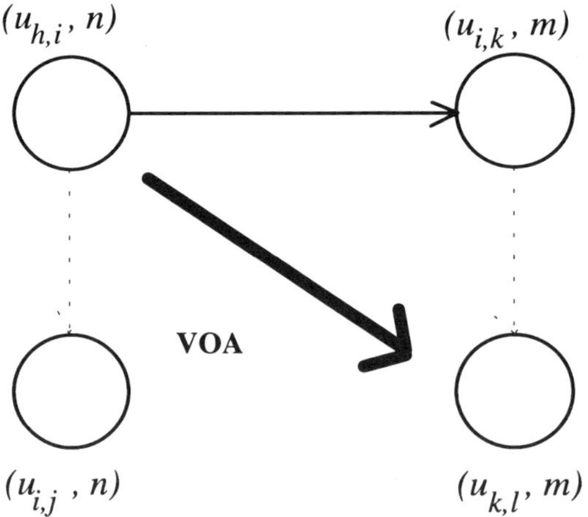

Figure 33 VOA on (S, GSO) vs. (S$_{gso}$, GSO$_{wds}$).

at t:

For S: $y_h(t) = 1$ and $u_{h,i}(t) = 1$.

For S_{gso}: $(u_{h,i}{}^\circ, n)(t) = 1$ and $w_{h,i;i,k}{}^{n,m}(t) = 1$

at t':

For S: $y_k(t') = 1$ and GSO_{sync} completes $u_{k,l}(t') = 1$ on T_m.

For S_{gso}: $(u_{k,l}{}^\circ, m)(t') = 1$ i.e. $((u_{k,l}{}^\circ, m),\_) \in K_- ((u_{i,k}{}^\circ, m), \_)$.

Q.E.D.

Comments. The reverse of the proposition 6:6 is not always true. If an ontological de-synchronization occurs at the S_{gso} state level, this does not necessarily correspond to an ontological de-synchronization in S. More precisely if an ontological de-synchronization occurs at the S_{gso} level such that a transition occurs from $((u_{h,i}{}^\circ,n), \_)$ to $((u_{k,l}{}^\circ, m), \_)$, then at the state set level of S, a de-synchronization can occur from $(y_h, u_{h,i})$ to any state whose control action is in $\Sigma_m(u_{g,g}{}^\circ) - \Sigma_m(u_{h,i}{}^\circ)$, where $u_{g,g}{}^\circ$ is the last action symbol in T_m. This result is expected since that is the reason why we cannot detect VOA on the non well determined state set S.

6.6.2 Unexpected External Action De-Synchronization

The property 4:6 states that for a well-determined state set S, an unexpected external action has a consecutive that is initial to a goal path. This property holds even for S_{gso}.

Proposition 6:7. Each unexpected external action in (S, GSO) is an unexpected external action in (S_{gso}, GSO_{wds}) and leads to a consecutive GSO-state which is initial to a GSO-goal path.

Proof.
According to the definition 6:10 of a control action for a GSO-controller, each GSO-control action in a GSO-goal paths is mapped to an expected external action at the object level. Therefore all the GSO-states that are initial to a GSO-goal are those which are <u>not</u> due to an expected external action at the object level.

Q.E.D.

6.6.3 Ill-Represented Formula De-Synchronization

Proposition 6:8. An ill-represented formula de-synchronization in (S, GSO), is in (S_{gso}, GSO_{wds}) one of the following :

> *(i) an unexpected external action de-synchronization, or*
> *(ii) an ill-represented formula de-synchronization.*

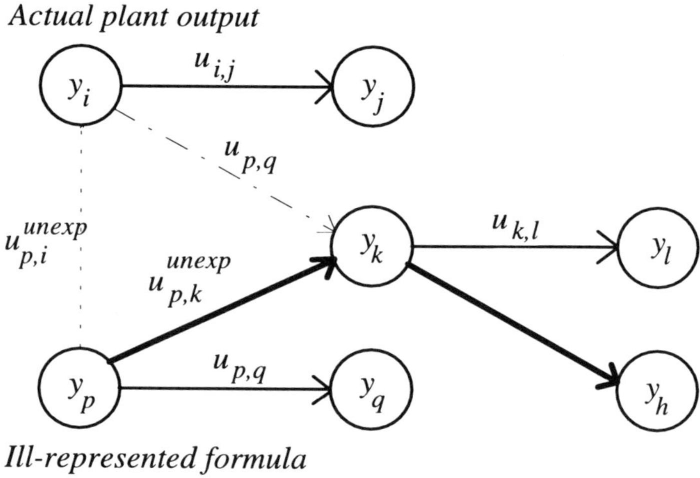

Figure 34 Ill-represented formula de-synchronization.

Proof.
The proof can be followed in **Figure 34**. The dash-dotted line show the actual control action and the bold lines show de-synchronizations.

For the *(S, GSO)* controller, let the state materialized at t be $(y_p, u_{p,q}) \in$ S and let y_p be ill-defined. That means, there exists another precondition formula y_i which is the actual plant output at t. Since, y_i is the actual plant output, we have that at t the non well defined state $(y_i, u_{p,q})$ materializes. (The state $(y_i, u_{p,q})$ is not well defined because $i \neq p$). Since this state is not well-defined, the postcondition y_q cannot materialize. Let us assume that at $t' > t$, some y_k materializes instead of y_q. According to the hypothesis, there exists a de-synchronization due to the ill-defined state at the object control level. That means there is no control action $u_{p,k}$ or expected external action $u_{p,k}{}^{ext}$. Let y_k be the precondition of some state $(y_k, u_{k,l})$ such that at t' the GSO completes the control action $u_{k,l}$. There are two cases: one when the postcondition of $u_{k,l}$ materializes and another one when the postcondition of $u_{k,l}$ does not materialize.

(i) Let us assume that the postcondition y_l of $(y_k, u_{k,l})$ materializes at some $t'' > t'$. At the object control level, that appears as if an unexpected external action $u_{p,k}^{unexp}$ has occurred followed by the control action $u_{k,l}$ which materialize its postcondition. The existence of the unexpected control action is deduced from the fact that $y_p(t)=1$ followed by $y_k(t')=1$ and that there is no expected external action $u_{p,k}^{ext}$ or control action $u_{p,k}$. Then, according to the proposition 6:7, the unexpected external action at the (S, GSO) level is an unexpected external action at the (S_{gso}, GSO_{wds}) level. That completes the proof for the first part of the proposition.

(ii) Let us assume that the postcondition y_l of the state $(y_k, u_{k,l})$ does not materialize as expected but another plant formula, y_h, materializes instead following a new de-synchronization. That means there are no control actions $u_{k,h}$ or expected external actions $u_{k,h}^{ext}$ in none of the states in S. For the (S_{gso}, GSO_{wds}) controller, the last two transitions translate either into a GSO-control action $w_{p,k;k,h}^{n,m}$ if a commutation can be defined from T_n to T_m or a GSO-expected external action $w_{p,k;k,h}^{ext}$ othewise. However, no GSO-state in S_{gso} has $w_{p,k;k,h}^{n,m}$ or $w_{p,k;k,h}^{ext}$ since there are no control actions with indices p,k and k,h in S which can appear in some T_n or T_m. Therefore the GSO-state with $w_{p,k;k,h}^{n,m}$ or $w_{p,k;k,h}^{ext}$ does not belong to the well-determined state set S_{gso}. That is exactly the characterisation of an ill-represented state for a well-determined state set as shown by proposition 5:3.
Q.E.D.

Remark: The ill-represented formula is translated correctly into an unexpected external action at the GSO-state level in case (i) due to the following: if the postcondition y_l of the control action $u_{k,l}$ materializes as expected, that means the control action $u_{p,q}$ is actually a possible control action that should appear as $u_{i,k}$ in the state set S (shown with dash-dotted line in **Figure 34**). In other words, $u_{p,q}$ and $u_{i,k}$ are the same physical action. In this case, the GSO-controller determines correctly the unexpected external action $u_{p,i}^{unexp}$, i.e. for the GSO-controller the causes for the existence of the ill-represented formula appear correctly as an unexpected external action that transforms the ill-represented formula into the actual formula.

6.6.4 Timing De-Synchronization

As shown in Section 5.3.4, a timing de-synchronization at the object control level may result in an ontological de-synchronization. Since, as shown by proposition 6:6, an ontological de-synchronization is preserved by the state transformation from S to S_{gso}, a timing de-synchronization at the object control level that occurs as an ontological de-synchronization, will result in an ontological de-synchronization at the GSO-control level as well.

6.7 Conclusions

In this chapter we have presented a state transformation which, under certain constraints, can transform a non-well determined state set into a well-determined state set. A controller can act with the transformed well-determined state set identically as with the initial one for a control without de-synchronizations except those due to expected external actions. Moreover the new state set has a specific behavior for de-synchronizations due to VOA, unexpected external actions, timing, and ill-represented formulas.

A question that may be raised relative to the use of sequences for GSO-states is about the relation among the object control configurations and the GSO-configurations represented as sequence indices. The definition of a well-determined state set is based on a number of properties relating controller configurations with control actions as described in Chapter 4. These properties are sound from a process control perspective. Thus one may expect that for a carefully designed controller these properties hold even if the configuration is not explicitly represented in the state set of the object controller. Then the question is the following: is it possible that at the GSO-control level the state set is well-determined but at the object control level the properties described in Chapter 4 relating control actions and configurations do not hold? For example, is it possible that S_{gso} is well-determined, the GSO-controller can detect all the four de-synchronizations by their distinct behavior, still, for the object controller there are cases when for instance control actions have the same effect upon the plant as external actions do?

The answer that follows from the proposition 6:5 is the following. Since at the object control level there is no available any information about the object controller configuration, the proposition 6:5 cannot guarantee that the state transitions at object control level have the properties described in Chapter 4. However, those properties of the object state set that lead to well-determined state set at the GSO-level are implicit in the order of goal paths and order of sequences the GSO uses. What follows from proposition 6:5 is that although there is no evidence that the properties from Chapter 4 hold at the object control level, there is no evidence either that these properties do no hold, inasmuch as this evidence can be read from the behavior of the controller during a de-synchronization.

7

THE ONTOLOGICAL CONTROLLER

Using the results obtained until now, in this chapter we can present a control architecture that can detect violations of ontological assumptions. This control architecture is intended for a controller with a well-determined state set. The well-determined state set is obtained either by design or after the state transformation described in Chapter 6. Either way, we call it as the *object controller* and we make no distinction between the two cases.

To be useful, the control architecture is required to have the following three essential properties:

- In cases when there are no violations of ontological assumptions, the control architecture allows a control that is identical to the one performed by the object controller.

- In cases of violations of ontological assumptions, the architecture ensures that a single state, called $S_{voa}{}^{oc}$ materializes, a state that indicates the existence of a VOA.

- The control architecture ensures that de-synchronizations due to VOA can be distinguished from de-synchronizations due to unexpected external actions, ill-represented formulas and timing.

First we show that $S_{voa}{}^{oc}$ cannot belong to the well-determined state set S of the object controller. Then we define a state set of the ontological controller, state set which includes $S_{voa}{}^{oc}$. Final conclusions and some ideas for future research concludes the book.

7.1 Motivations For The Ontological Control Architecture

The proposition 4.1 shows that a PC with a well-determined state set can exhibit violations of ontological assumptions by means of a specific de-synchronization into the set of collateral states of the state with VOA. The state set of a PC is either well-determined by design, or it can be transformed into a well-determined state set with the state transformation described in Chapter 6.

Although the PC with a well-determined state set does have a specific behavior in cases of VOA, the PC is not "aware" of this behavior. Thus it will continue to control according to its state set at the object control level even under VOA. As already shown in Chapter 5, this control is not the intended one and the PC will not materialize any of its goal states. Consequently we need a state, $S_{voa}{}^{oc} = (y_{voa}{}^{oc}, -)$ such that $y_{voa}{}^{oc}$ interprets to 1 if and only if there is a violation of the ontological assumptions. One may assume that this state may be already in S or it may be a new one, added to S. However, the following proposition shows that $S_{voa}{}^{oc}$ cannot be in S, nor added to S.

Proposition 7:1. Let S be a well-determined state set and $S_{voa}{}^{oc}$ be a state which materializes in cases of a VOA on any state of S. Then $S_{voa}{}^{oc}$ is not well-defined.

Proof.

Let $S_p{}^a$, $S_q{}^b$, $S_n{}^c$, $S_m{}^d$ be four states in S such that $S_p{}^a \rightarrow S_q{}^b$ and $S_n{}^c \rightarrow S_m{}^d$. According to the hypothesis, the state $S_{voa}{}^{oc}$ materializes after a VOA on any state of S. Let two such states with VOA be $S_q{}^b$ and $S_m{}^d$. According to the hypothesis, if a VOA occurs on $S_q{}^b$ we have a transition $S_p{}^a$ to $S_{voa}{}^{oc}$ and if a VOA occur on $S_m{}^d$ we have a transition from $S_n{}^c$ to $S_{voa}{}^{oc}$. Note that these transitions cannot be state transitions since it is not known in advance when a VOA will occur. Therefore all the transitions to $S_{voa}{}^{oc}$ are due to external actions. The transitions due to external actions do not change the configurations. We use a consequence of the proposition 5:1 as follows. A VOA on $S_p{}^a$ means that the configuration c^b of the state $S_q{}^b$ is already realized at the time instance when the ontological de-

synchronization occurs from $S_p{}^a$ to $K_-(S_q{}^b)$. Therefore, the state $S_{voa}{}^{oc}$ must have the configuration c^b. If not, the precondition of $S_{voa}{}^{oc}$ cannot materialize. The same argument for the VOA on $S_m{}^d$ requires that the configuration of the state $S_{voa}{}^{oc}$ is c^d. However, the configuration formulas are not subsumed (see Section 4.1.2), therefore the state $S_{voa}{}^{oc}$ cannot have a configuration like $c^b \vee c^d$. Thus $S_{voa}{}^{oc}$ is not well-defined.
Q.E.D.

The control meaning of the proposition 7:1 above is that a VOA results in certain <u>type</u> of state de-synchronization (i.e. the *ontological de-synchronization*). The state set S is not at the abstraction level to represent this type of de-synchronization.

A consequence of the proposition 6.1 is therefore that the ontological controller requires a state set at a different abstraction level than S. The preconditions of the ontological controller are in terms of types of state transitions. We proceed in the following section with the definition of these preconditions.

7.2 The State Set Of The Ontological Controller

The state set of the ontological controller consists of pairs <precondition, action>, identically to any other type of controller. We shall refer to the components of an ontological controller as oc-states, oc-preconditions and oc-actions. We distinguish the oc-states, oc-preconditions and oc-control action state notation from the object controller with a superscript 'oc'.

The states can be determined from Chapter 5, (summary in Table 3). Each type of transition or de-synchronization has a corresponding state. Following Table 3, we distinguish the following states of the GSO-controller:

- $S_{cntr}{}^{oc} = (y_{cntr}{}^{oc}, u_{cntr}{}^{oc})$ is the oc-state materialized for a transition due to an expected external action or a state transition due to a control action at the object control level.

- $S_{unexp}{}^{oc} = (y_{unexp}{}^{oc}, u_{unexp}{}^{oc})$ is the oc-state materialized for a de-synchronization due to an unexpected external action at the object control level.

- $S_{voa}{}^{oc} = (y_{voa}{}^{oc}, u_{voa}{}^{oc})$ is the oc-state materialized for an ontological de-synchronization at the object control level.

- $S_{illrepr}{}^{oc} = (y_{illrepr}{}^{oc}, u_{illrepr}{}^{oc})$ is the oc-state materialized for a de-synchronization due to an ill-represented state at the object control level.

Thus the state set of the ontological controller consists of only four states as follows:

$$S^{oc} = \{S_{cntr}{}^{oc}, S_{unexp}{}^{oc}, S_{voa}{}^{oc}, S_{illrepr}{}^{oc}\}$$

7.2.1 The OC - State Precondition

In this section we consider a PC with a well-determined state set. We use the simpler notation for the states of the object controller, i.e. $S_i = (y_i, u_{i,j})$. We assume that y_i consists on a pair of proper plant formula (z_i or $u_{i,j}{}^{\circ}$) and configuration ($c_i{}^p$ or n) and the control action is some $u_{i,j}{}^{p,q}$ or $w_{i,j;j,l}{}^{n,m}$ (the choice for the first or second notation depends on whether the state set is well-determined from the beginning or after the transformation described in Chapter 6).

Let $S_i = (y_i, u_{i,j})$ be the current state of the object controller materialized at t-n and $S_j = (y_j, u_{j,l})$ be the consecutive state to S_i: $S_i \rightarrow S_j$. Let the precondition formula materialized at some $t > t$-n be y_n. Since the state set S is well-determined, there exists a single state S_n in S such that $S_n = (y_n, u_{n,m})$. The inputs for the interpretation of the precondition formulas of the ontological controller is the pair $\{u_{i,j}, y_n\}$.

Definition 6.1. A control action $u_{i,j}$ of a state S_i and a precondition formula y_n of a state S_n (S_i , $S_n \in S$ and S is well-determined) are inputs for the precondition formulas of the ontological controller iff $u_{i,j}(t\text{-}n)=1$ and at t the plant formula that materializes after the execution of the action $u_{i,j}$ is y_n ($y_n(t)=1$).

In other words, the symbols in the pair $\{u_{i,j}, y_n\}$ are *plant signals* in X for the ontological controller (see Section 3.1.4). With these two symbols, the plant formulas can be determined as follows.

- The precondition formula for the oc-state materialized in cases when a consecutive state materializes or an expected external action occurs. If the consecutive state to S_i materializes, then $n \equiv j$, that is, the input pair is actually $(u_{i,j}, y_j)$. If, alternatively, an expected external action occurs, then S_n is a state reachable from S_i by an expected external action, that is, S_n is a state in $K_-(S_i)$. These two condition together give the definition for $y_{cntr}{}^{oc}$:

$$y_{cntr}^{oc} = ((S_n \in K_-(S_i) \wedge u_{i,n}^{ext} \in U^{ext}) \vee (n \equiv j))$$

- The precondition formula for the oc-state materialized for an unexpected external action. As shown in Section 5.3.2, an unexpected external action materializes a state in $K_-(S_i)$ identically to an expected external action, but there is no symbol for the corresponding external action:

$$y_{unexp}^{oc} = (S_n \in K_-(S_i) \wedge u_{i,n}^{ext} \notin U^{ext})$$

- The precondition formula for the oc-state materialized for violations of ontological assumptions. According to proposition 5:1, for an ontological de-synchronization, the precondition formula y_n belongs to a state in $K_-(S_j)$:

$$y_{voa}^{oc} = S_n \in K_-(S_j)$$

- The precondition formula for the oc-state materialized for ill-represented states. The results in Section 5.3.3 show that the precondition formula materialized after a de-synchronization due to an ill-determined state cannot be in the collateral of S_i, S_j and cannot be a consecutive to S_i:

$$y_{illrepr}^{oc} = (n \neq j) \wedge S_n \notin K_-(S_j) \wedge S_n \notin K_-(S_i)$$

7.2.2 The OC- State control action

The oc-control actions are recovery operations at the level of the object controller. The following cases result from the architecture:

- The oc-control action u_{cntr}^{oc} for the case when the consecutive state materializes or an expected external action occurs, is simply to allow the control to proceed since no problematic control situation was detected.

- The oc-control actions for ill-represented states ($u_{illrepr}^{oc}$) and unexpected external actions (u_{unexp}^{oc}) are not in the scope of ontological control. Although of considerable value in applications, the detection of ill-represented states and unexpected external actions was only required for the proof that using a well-determined state set we can distinguish among different causes for de-synchronizations. The effective control actions that have to be executed when these problematic control situations occur, are not in the scope of this book.

- Relative to the oc-control action for cases of violations of ontological assumptions, $u_{voa}{}^{oc}$, the results in Chapter 5 give the following limitations:

  - In the book we have assumed that the state set S is fixed once and forever. The results in Chapter 5 show that if S is well-determined, then no goal state can be reached after a VOA, control actions do not have expected effects and the control continues in a cycle consisting on initial states of goal paths. Therefore, under the assumption that the state set is fixed and it is well-determined, the only appropriate oc-control action is to stop the controller.

  - If the restriction that the state set S is fixed is lifted, then after the occurrence of a VOA there may be possible to design a new state set that has different control actions, preconditions and configurations such that the control with the new state set has no violations of ontological assumptions. Since the state with VOA is known, the new state set can be designed such that the assumptions not considered for the state with VOA appear explicit in the new state set. Since this operation is performed at the control knowledge level (see Section 3.3), the details of this operation are not in the scope of this book.

  - An application-independent method has the following limitations. Since the state set is complete, there are no more states that can be add to the existing state set. However, the architecture indicates which is the culprit state. If the culprit state is for instance S_i, then S_i cannot materialize in the presence of violations of the ontological assumptions. That means the proper plant formula part of S_i it is not bound right and thus S_i can be removed from the state set. However, the states in $K_-(S_i)$ are also not bound right since states in $K_-(S_i)$ materialize when S_i does not materialize. That means the plant formulas in S_i use the same sensor signals as those in $K_-(S_i)$ but with a combination of complementary intervals. Thus both the state Si and the state set $K_-(S_i)$ can be removed and new states can be created instead. The new states must have the same configuration part in the plant formula but a different proper plant formula. Now we identify two cases. (i) If the proper plant formula of the culprit state has only discrete variables (e.g., only digital inputs), then the recovery is not possible and the object controller should stop. The reason is that there cannot be created new states since the initial state set is complete and all the conbinations of discrete values are already there. (ii) If the proper plant formula has continuous parts, such as analog signals which are in certain intervals, then new states can be created automatically by using new intervals and a specification of the "power" of the control action of the culprit formula. Essentially this is a fuzzy logic approach. Some methods are decribed in [], [], [].

Summary. The following table summarizes the state set of the ontological controller:

TABLE 5: The state set of the ontological controller

OC-State	Meaning
S_{cntr}^{oc}	Object states materialized due to control actions or expected external actions
S_{unexp}^{oc}	It has occurred a de-synchronization due to unexpected external action
S_{voa}^{oc}	It has occurred a de-synchronization due to a violation of ontological assumptions
$S_{illrepr}^{oc}$	It has occurred a de-synchronization due to an ill-represented formula

7.3 The Architecture Of The Ontological Controller

The precondition formulas for the ontological controller states use the pair $(u_{i,j}, y_k)$ consisting on a control action $u_{i,j}$ materialized at certain time and the plant formula y_k of a state S_k that materializes consecutively, following the execution of $u_{i,j}$. The architecture of the ontological controller makes available this pair as the input of the ontological controller. In cases when a violation of the ontological assumptions is detected, the oc-control action is an input to the PC that either stops the PC or perhaps changes its state set. The resulting control architecture, based on the **Figure 9** (for initially well-determined state sets) or **Figure 26** (for transformed state sets) is shown in **Figure 35**. The *"Delay"* block stores the last control action symbol for a one state transition period such that the control action $u_{i,j}$ has a time stamp prior to y_k as required by Definition 7:1.

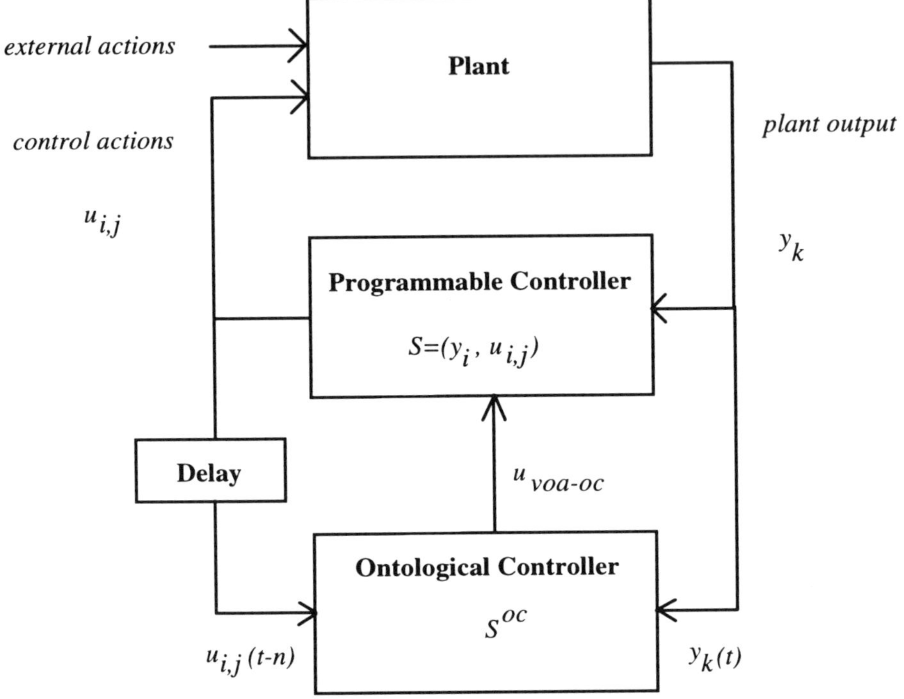

Figure 35 Ontological Control Architecture.

7.4 Conclusions

In short, the results reported in this book are the following.

- Chapter 3 introduces informally and Chapter 5 formally a definition for the concept of *violations of ontological assumptions* (VOA). Violations of ontological assumptions are shown to produce state de-synchronizations. However, for an unrestricted state set, the de-synchronizations due to VOA cannot be distinguished from other causes for de-synchronizations. In the subsequent chapters we give an answer to this problem.

- Chapter 4 introduces a number of restrictions on the state set S of a controller. A state set that conforms to these restrictions is called *well-determined*. In Chapter 5 we show that for well-determined state sets, VOA has a specific, recognizable behavior that is distinct from other causes of de-synchronizations.

- In Chapter 6 we describe why in applications the state set of a PC may not be well-determined. A state transformation is presented which, under certain constraints, can transform a state set that is not well-determined into a state set that is well-determined.

- Finally, Chapter 7 shows an architecture that is built on the object controller. The ontological control level of it can materialize a control state at the ontological control level when violations of the ontological assumptions occur at the object control level. The same architecture can distinguish other causes of de-synchronizations that occur at the object control level such as ill-represented formula, unexpected external actions and timing.

7.5 Future Research

This book gives one solution for the problem of control under violations of ontological assumptions. However, there is no proof that this solution is unique. There may be also other solutions to this problem. These solutions may be found by further research in the following areas:

- Advanced algorithms for the synchronization operation and the goal seeking operation. A more careful investigation of these two operations reveals that the difficulty is to find the right priority between which one of these two operations is allowed to act when de-synchronizations occur due to problematic control situations.

- We have based our solution for a well-determined state set on the concept of control configuration. However, there may be other concepts not based on configuration which can create state sets which reveal violations of ontological assumptions.

- There may be more state transformation operations, with less constraints that the one presented in this book, which can transform a non-well-determined state set into a well-determined state set.

- If the state interpretation is not done with crisp $\{0,1\}$ values but with fuzzy intervals, then the state set may be changed after a VOA by associating fuzzy rules to states such that these rules create a state set which has no VOA. This approach is one of the most promising solutions today for the control action of the ontological controller.

REFERENCES

[1]   D. Driankov, G. Fodor, "Fuzzy control under violations of ontological assumptions, Invited plenary talk, FLAMOC'96 Proceedings, pp. 109-115, Sydney, Australia, Jan. 15-18, 1996.

[2]   J.L. Grantner, G. Fodor, D. Driankov, M.J. Patyra Application of the Fuzzy State Fuzzy Output Finite State Machine to the Problem of Recovery from Violations of Ontological Assumptions. in Intelligent Enginering Systems vol. 6 , pp. 277-282, ASME Press, 1996.

[3]   G. Fodor, D. Driankov A New Approach to On-line Fault Identification in PLC Control. To appear in Proc of the SAFEPROCESS'97 Conference, Elsevier Science Ltd., Oxford, 1997.

INDEX